THE MANAGEMENT OF RISK TO SOCIETY FROM POTENTIAL ACCIDENTS

THE MANAGEMENT OF RISK TO SOCIETY FROM POTENTIAL ACCIDENTS

The main report of the UKAEA Working Group on the Risks to Society from Potential Major Accidents, with an Executive Summary edited by

F. R. Allen
A. R. Garlick
M. R. Hayns
A. R. Taig

SRD, AEA Technology, Culcheth, Cheshire, UK

CRC Press
Taylor & Francis Group
Boca Raton London New York

CRC Press is an imprint of the
Taylor & Francis Group, an **informa** business

First published 1992 by CRC Press
Taylor & Francis Group
6000 Broken Sound Parkway NW, Suite 300
Boca Raton, FL 33487-2742

Reissued 2018 by CRC Press

British Library Cataloguing in Publication Data

Management of Risk to Society from
Potential Accidents
I. Allen, F. R.
363.1

ISBN 1-85166-892-6

Library of Congress CIP data applied for

A Library of Congress record exists under LC control number: 92026262

Publisher's Note
The publisher has gone to great lengths to ensure the quality of this reprint but points out that some imperfections in the original copies may be apparent.

ISBN 13: 978-1-315-89807-0 (hbk)
ISBN 13: 978-1-351-07717-0 (ebk)

Visit the Taylor & Francis Web site at http://www.taylorandfrancis.com and the
CRC Press Web site at http://www.crcpress.com

THE MANAGEMENT OF RISK TO SOCIETY FROM POTENTIAL ACCIDENTS

Executive Summary

F R Allen
A R Garlick
M R Hayns
A R Taig

CONTENTS
(Executive Summary)

1 INTRODUCTION

1.1 Background

The study by AEA Technology of the technical basis for the management of risks to society from potential major accidents, summarised here, was initiated after the occurrence of two significant events. The first of these was the accident at the Chernobyl nuclear reactor site in the Ukraine in April 1986. This, the most serious nuclear accident experienced, illustrated the impact which a major accident can have on the country involved and also in surrounding countries. This impact involved a number of aspects and these did not entirely coincide with those previously considered in the context of "societal risk" by risk analysts. "Societal risk" refers to the risk to society as a whole and may also be termed "social risk".

The second event was the publication in January 1987 of Sir Frank Layfield's report on the Public Inquiry into the CEGB's proposal to build a PWR at Sizewell. Layfield's report (1987) contains a number of conclusions and recommendations relating to the analysis and assessment of risk, including societal risk, and to the way decisions involving these risks are made in the UK. He accepted that the levels of individual risk to members of the public corresponding to the CEGB's Design Safety Criteria (CEGB, 1987) could be tolerated "providing that there is expected to be economic benefit sufficient to justify the risks incurred". However he was unable to reach any conclusion with regard to what he called "social risk".

AEA Technology initiated this study to support its own decision making on risk management of the nuclear plants and laboratories it controls. But the principles underlying decisions on social risk are of much broader applicability. Since the study was completed they have become topical in many other sectors of the transport, process and energy industries, and through publication of this work we hope to provide a source document of widespread usefulness.

1.2 AEA Technology Policy and the Aims of WGRSPMA

AEA Technology is responsible for the safety of operations on its own sites. It also carries out research into the safety of nuclear plant on behalf of the nuclear industry as a whole and so as to be able to advise the UK Government on such matters. Within AEA Technology, SRD, (AEA's safety and reliability consultancy business) acts both as a source of independent advice on the safety of operations and as a centre for safety research, development and consultancy to industry at large. SRD has a long-standing interest in the problems of risk assessment and risk acceptability, going back some 25 years to when F R Farmer, FRS, then Director of SRD, published a safety target closely related to societal risk. This target remains one of the few societal safety targets which have been seriously applied.

The control of social risk and its reflection in specific risk guidelines or targets is a complex matter, and AEA Technology view its current position as one which, though established after long and serious consideration, should continue to evolve to reflect the changing role of the organisation and the social and political environment in which it operates.

AEA Technology's policy on risk management has evolved over many years and is currently embodied in a Corporate Policy and Safety Directives document (AEA Technology, 1991), the basis of which on radiological risk matters is more fully explained in the associated Code of Practice and guidance notes on the control of radiological hazards (UKAEA, 1987). From 1990, AEA Technology has been subject to the same requirements for licensing of nuclear sites by HM Nuclear Installations Inspectorate as has the commercial nuclear industry in the UK. Thus current AEA Technology risk management policy and guidelines also reflect the current UK licensing requirements for nuclear plant generally.

Risk targets are specified for individuals most at risk from AEA Technology's activities - both staff and members of the public, and also for the maximum tolerable frequency of large accidents. It is this latter target which, through a very simple risk measure, provides the basis for control and management of a wide variety of hazards and concerns associated with large radioactivity releases. This is the current compromise which the UK nuclear industry has struck between explicit guidelines for protection against the aggregated impact of different risks on society, and useful guidance for designers, operators and management of plant which can readily be applied to parameters within their control.

The Working Group on the Risks to Society from Potential Major Accidents (WGRSPMA) was set up in support of the development of AEA Technology corporate policy in the specific area of societal risk. It carried out the bulk of this work during 1988 and 1989. Its chairman and secretariat were SRD based but its membership included safety experts from throughout AEA Technology and from other bodies.

The Working Group decided that a review of technical issues related to societal risk would be useful both within the nuclear industry and more generally. This is that review; its objectives are to:

- provide the essential background for the interested non-expert, particularly via the references
- outline the major problems in the field
- report on work carried out by the Group to throw light on some of these areas, and
- make general recommendations on matters such as definitions, methods for estimating and representing risk, principles for risk acceptability and means of making decisions.

Because of its background as part of AEA Technology's policy formulation process, this document, while considering risk issues in general, makes particular reference to their relevance to the nuclear industry. But it is not an objective of this document to formulate specific quantitative safety goals for AEA Technology or any other body.

Since completion of this study, AEA Technology has become the trading name of the United Kingdom Atomic Energy Authority (UKAEA), and the names UKAEA and AEA Technology are both used in this document. The UK electricity supply industry has been privatised, and the nuclear operations of the former CEGB are now carried out by Nuclear Electric; CEGB is more often referred to in this document, particularly in the context of the case made and extensively analysed at the Sizewell 'B' enquiry.

1.3 Study Report

This Executive Summary document summarises the main WGRSPMA report (UKAEA, 1989) published as a result of the study mentioned above; the Executive Summary is intended to be a standalone summary of the results of the work but readers requiring more information are referred to the main report.

2 RISK

2.1 The Nature of the Risk

Risk is a widely acknowledged concept, but one whose exact meaning is evolving. It now embraces both individual risks and also risks to society at large and to the environment. Further, we now recognise the perception of risk as an important and variable factor. Life in modern society is becoming, statistically, less risky when measured in terms of life expectancy (BMA, 1987). Despite this, increased awareness of hazards, partly as a result of continuing scientific investigation which bring such hazards to the public's notice, leads to the widely-held public belief that life is getting riskier. This indicates that, besides statistics, many other factors enter into the public's perception of risk, such as: the potential for catastrophe; lack of familiarity with the risk; the involuntary nature of the risk; scientific uncertainty; lack of personal control; risk to future generations; doubtful benefits; inequitable distribution of risks and benefits; and potentially irreversible effects. In addition, perception of risk depends on local cultural values.

Many definitions of risk have been proposed including those of the Royal Society Study Group (Royal Society, 1983) and the Institution of Chemical Engineers (1985). The definitions adopted in the report closely follow the latter:

> "Hazard: a physical situation with a potential for human injury, damage to property, damage to the environment or some combination of these.
>
> Risk: the likelihood of specified undesired events occurring within a specified period or in specified circumstances arising from the realisation of a specified hazard. It may be expressed as either a frequency (the expected number of specified events occurrin, unit time) or a probability (the probability of a specified event following a prior event) depending on the circumstances.

This definition contains the essential elements of all definitions of risk: that of carefully defined events, with associated probabilities (or frequencies). Frequencies can be interpreted in various ways. Mathematically they may be expressed as fractions of an event, often very small fractions, eg "10^{-6} events per year". Though a mathematically valid concept, this representation is perhaps less clear to the non-expert than the equivalent forms such as "1 event in 1 million years" or "a 1 in a million chance it will happen this year". Alternatively we might say that this probability of the event in a year is 10^{-6} (or 1 in a million).

2.2 Societal Risk

The above definition is very general. It can clearly apply to risk to individuals ie "individual risk". It can also be applied to recognisable groups, such as plant operators. Individual risk criteria might take the form of limiting the risk of death to any member of the public to a target of 10^{-7} per year (ie 1 chance occurrence in 10 million years). Such targets have been used to give advice to designers and operators of plant on more detailed plant-specific requirements such as the reliability needed for a particular system (see for example Brown 1986).

It is concluded (UKAEA 1989 Section 2.4.1) that the use of such criteria alone could be criticised, however, in that they might miss important factors relating to the total effects of a release; if a large number of people were exposed to a risk which was, individually, sufficiently small to meet the individual risk criterion, the total effect on the group as a whole might nevertheless be unacceptably high. For example, an accident may meet the individual risk target of 10^{-7} per year quoted above, but may nonetheless result in 100 deaths if the exposed population is high enough. This might perhaps be considered unacceptable even though the individual risk level could be tolerated.

The main study report points out (UKAEA 1989 Section 2.4) another aspect omitted from simple individual risk criteria which is the consideration of a wider range of consequences, such as temporary or permanent evacuation of large numbers of the public loss of crops, etc. Thus, society as a whole is concerned about a number of aspects of risk in addition to the risk to individuals. In addition, society as a whole is the recipient of the benefits from technological activities which are balanced against the risks involved.

Therefore the balancing of societal risks and benefits is different from that for individuals, and there is no a priori reason why these two activities (viz societal risk evaluation and individual risk evaluation) should always come to the same conclusions. Targets for societal risk may thus impose requirements on plant designers and operators which are different from those for individual risk.

2.3 Characterising and Representing Risk

Types of risk may be specified in terms of the types of harm arising from the postulated undesired events. Thus risks may include:

Personal Risks:- Risks affecting individuals, such as early death, later death (eg due to cancer or other disease), serious incapacity, minor injury, forced permanent or long-term evacuation, serious birth defects.

Non-Personal Risks:- Risks to society as a whole might in addition include a wider range of harms, such as: population evacuation, resettlement, clean-up operations, interdiction of food, provision of uncontaminated water, sterilisation of land or property, replacement of lost capacity. Many of these can be represented as financial losses.

Societal risks, or social risks, are meant to include the risk to society as a whole. They therefore need to embrace the summation of personal and non-personal risks.

The ways such risks may be characterised are reviewed in Chapter 3 of the WGRSPMA report (UKAEA 1989). One approach is by producing weighted sums of these various aspects. This is not recommended however since we do not consider it appropriate to compare unlike effects in this way. Another means would be to weight such aspects in terms of the costs involved ie to produce the total costs of an accident in monetary terms. Such a course, however, requires quantitative value to be placed on loss of life, which is fraught with difficulties.

Risks associated with radiation exposure pose additional difficulties due to the stochastic nature of the effects at low levels of exposure. At these levels the concern is the possible development of cancer. This may occur 10 to 30 years after the exposure; different types of cancer may be induced; the probability of developing cancer may vary with age or sex, and may be affected by other factors such as occupation. Thus the risk to an individual is variable because of a number of factors, and whether cancer develops or not will in addition be uncertain in a purely statistical sense. The risk to society can be evaluated by summing the risks to individuals to produce the predicted number of deaths which, assuming perhaps pessimistically that individual risk is proportional to radiation dose (the "linear hypothesis"), gives a predicted number of deaths proportional to the total (or collective) dose to the population. An alternative approach would be to sum the loss of life expectancy, which would take account of the considerable delay in developing cancer after low doses of radiation.

As noted above, the predicted number of fatalities following a release of radiation is proportional to the collective radiation dose and thus to the release. However, other harms such as evacuation, interdiction of food, etc, vary in a more complex way with the magnitude of the release, showing threshold effects etc. Furthermore, all risks are likely to be dependent on the specific site concerned.

Thus it is concluded (UKAEA 1989 Section 3.3.2) that general targets relating to the safety of the public should not be set in terms of release, although it may well be necessary to interpret risk targets in terms of releases from specific sites.

Societal risks may be presented in a number of forms, graphical, tabular and functional. In contrast, individual risks are simply numbers and so are adequately represented by tables. Graphical representations include a number of approaches. Frequency distributions show the frequency of a given level of harm; these may be shown by histograms (which are fairly straightforward) or scatter plots (which require more careful definition). Frequency densities may be plotted but again careful definition is required and this presentation may be confused with others. Complementary Cumulative Distribution Functions (CCDF's) are technically the most satisfactory representation. In this representation, at each value of consequence (C) is plotted the frequency of events with consequences equal to or greater than C. (This is also commonly called an f/N line). CCDF's are recommended for general use; however we note that the Layfield report opposes the use of CCDF's on grounds of difficulty of interpretation. Even though the main report endorses the CCDF from a 'professional' risk assessment viewpoint, it recognises the need for the analyst to provide simpler, though less rigorous, interpretations. A further representation is the first moment cumulative distribution; this has some technical advantages but is considered too complex for wide application.

2.4 Existing Risks

No activity can be undertaken without some associated risk. There exists, therefore, a background of risks associated with everyday life, including the working environment, accidents of all kinds, leisure activities, and medical conditions. Our knowledge of these is reliant on historical data; these are only slowly being compiled, and are very uneven in their coverage. Other risks are not easily estimated from historical data; examples include cancer from background radiation, or cases where the risk is dominated by the unrecorded injuries not the recorded fatalities. Rare events also pose problems when relying on historical data.

Despite these difficulties, estimates of existing levels of both individual and societal risk have been made in the main WGRSPMA report (UKAEA 1989 Chapter 4) as follows:

Existing levels of average individual risk in Great Britain are 10^{-2} per year for death from all causes, but for the group least at risk (females aged 5-15) this falls to about 2×10^{-4} per year. The corresponding figures for death due to cancer are

3×10^{-3} per year and 4×10^{-5} per year respectively, and for accidents, 3×10^{-4} per year and 9×10^{-5} per year.

Risks of fatal injuries to those at work vary between 1×10^{-6} per year and 180×10^{-6} per year depending on the industry. Risks of injury at work vary between 1×10^{-4} per year and 28×10^{-4} per year, depending on the industry. The data indicate that these risks are varying with time and that the current trend is towards increasing safety. However it must also be pointed out that the individual risk statistics are very sensitive to the choice made by the analyst of the population at risk; for example when considering risks associated with an industry, these risks may well be unevenly shared between different types of worker within the industry, some being higher than the average and others lower. This effect might work to the disadvantage of the nuclear industry, as would the comparison of cancer risk with that of early death in other industries.

With regard to societal aspects, it is concluded that both natural disasters and accidents associated with man's activities have the potential for large consequences. These can be compared using CCDF's and a number of these are presented in the body of the report. Risks (viewed here as frequency x consequences) from many natural disasters seem to be biased towards the high-consequence events whereas risks from man's activities seem to be more evenly spread between high and low consequence events on a relative basis.

2.5 Estimating Risks

Risks are estimated, analytically or predictively, by a number of techniques (reviewed in UKAEA 1989 Chapter 5), which are collectively known as probabilistic safety assessment, PSA (or sometimes probabilistic risk assessment, PRA). These techniques aim to analyse the system, plant, or installation to predict the modes of failure of the system and their frequencies, and to model the progression of accidents to predict their consequences. Many of the techniques used are long established and there can be reasonable confidence in the results generated. However, a number of areas of uncertainty remain where the applicability of these techniques is questionable, for the moment at least.

The analysis of plant systems to predict failure modes and frequencies has evolved from reliability techniques developed over a number of decades in various industries. The logical model of the plant can give qualitative insights into strengths and weaknesses in the design as well as quantitative values of reliability. However the main thrust of its use in PSA is to identify plant failure states and to predict expected frequencies for these. These are based on reliability data for individual components built up over a number of years, together with other inputs. The uncertainty in the data is partly statistical, but additional uncertainties relate to its applicability to the system in question. Further uncertainties can arise also, however. Systems designed for high reliability achieved by in-depth back-up provisions may be vulnerable to so-called "common cause failures" (or "dependent failures") which result from unforeseen interactions or common features causing coincident failure of back-up systems. In addition the question of completeness arises - has a sufficiently wide range of failure modes been considered?

The prediction of plant consequences by physical modelling is a more recent development and is rather different in character. The application of models, even well developed and verified ones, to specific situations may require extrapolation, and for less studied phenomena the models may be very crude. Expert judgement may also be involved to a considerable extent, because of lack of well established models and data. Therefore, the uncertainties in this stage of the analysis are of a different kind than those for the systems analysis stage, and may in addition be significantly greater.

Human factors are also identified as a potential source of uncertainty. Human operators can intervene in a number of ways which may be beneficial or detrimental; a complete analysis of these is not currently possible although it is possible to analyse a plant's robustness against operator intervention in potentially detrimental ways. However, this is an element of risk which it is not possible to estimate quantitatively using current probabilistic methods, indeed it may not be possible at all, and so these aspects are currently considered separately from PSA analysis, and the results of these analyses must be treated with due caution.

Hazards are another area which is currently treated separately from the main PSA analysis. Hazards refer to events which impinge on a system from outside; they include fire, which can arise within the plant, or such events as aircraft crash and earthquakes, which arise outside the plant. These have a similar character to dependent failures in that they can cause coincident failure of several systems, eg main and back-up systems. As a result it is possible for major external events, such as large earthquakes, to dominate the risk from high reliability plant despite their very low probability of occurrence. This may then present a problem because of the difficulty of quantifying the probability of such rare events.

The WGRSPMA report concludes (UKAEA 1989 Section 5.5) that, because of these limitations, probabilistic methods cannot provide the complete safety case for hazardous installations and so more traditional 'deterministic' methods (in which safety is demonstrated using a prescriptive requirement on bounding cases to be treated and pessimistic methods to be used) will remain important also. In evaluating risk using PSA results, it is necessary to take into account the various uncertainties and subjective factors indicated above; these should be quantified as far as can reasonably be done. However, PSA methods have a great number of benefits from the qualitative viewpoint in generating deeper understanding of the overall plant design as well as in allowing quantification of risk, for which they are the only means available.

3 GOALS AND TARGETS

3.1 Acceptance and Tolerability

Section 6.2 of the WGRSPMA report (UKAEA 1989) analyses the reasons for risk acceptance and risk tolerance. Risks can be accepted either if they are not recognised as such or if they are recognised but perceived to be insignificant. Tolerance of a recognised, significant risk implies an appreciation of benefits which are seen to outweigh the risk. The basic difficulty is that risk is not '*measurable*' in a physical sense. It may be calculated, albeit imperfectly, but people's innate responses of '*fear*' and '*dread*' tend to impinge on what logically may be an otherwise technical issue. It is necessary therefore to identify, qualitatively, three levels of concern:

Unconcern may arise because of ignorance either of the hazard's existence or of its extent; for example the UK public's unconcern over the risk from high winds may have changed following the widespread gale damage in October 1987. Quite distinct from this, however, is the concept of a '*de minimis*' level of risk which is known to be so small as to be of no concern.

Concern or Reluctance may arise when there is either a greater awareness of a hazard whose magnitude is difficult to assess or an awareness of a higher level of risk from a hazard which can be assessed more confidently. Acceptance then rests on some substantial benefit which is perceived to outweigh the risk.

Wholehearted Acceptance for example by insurance companies is characterised by a high degree of confidence that the benefits will outweigh the risks. This confidence may rest on precise estimates of the risk or on a high level of benefit which outweighs the uncertainty in the risk estimate.

From these considerations, three points emerge - first the '*de minimis*' concept of a risk too small to bother about, second the need to take into account not only the assessed risk, but also the confidence with which it can be estimated, and thirdly the need to take into account public perception which includes many qualitative aspects of the risk involved.

It is concluded that successful risk management and the achievement of risk acceptance or tolerance depends either on demonstrating a risk below the '*de minimis*' level or on balancing risks and benefits. However, the subjective aspects of risk perception make this latter very difficult. Various possible strategies for this are reviewed in the main report (UKAEA 1989 Section 6.3):

Evaluation of Risks against Benefits

(i) Unique Benefits

It can be argued that some technologies offer unique benefits so that there is no particular reason to insist that their risks should be smaller than the risks of other activities. This is a reasonable (ie logical) strategy for risk management but agreement on the uniqueness of the benefits may be difficult to obtain.

(ii) Equivalent Benefits

In this approach, various technologies offering equivalent benefits would be compared; thus for example, nuclear power would be compared with other technologies for producing electricity so that risk management can be applied to minimise the overall risk. Again this is a reasonable (logical) strategy but the difficulty is the wide differences between the kinds of risks involved with the different technologies, and the differences in the perception of these risks.

Evaluation of Risks in Isolation from Benefits

(i) Too Small to Worry About

Risks below a certain '*de minimis*' level would be accepted as being too small to worry about. The problems here are associated with identifying such risks, and in making comparisons with technological risks such as that from nuclear power where psychological factors may affect the perceived risk comparison quite considerably.

(ii) Too High to Tolerate

It would certainly be useful to identify a level of risk which was too high to tolerate under any circumstances. This is not considered further here, but this has been considered by HSE in their recently-published document on the tolerability of risks (HSE 1988).

3.2 Risk Comparisons

Since tolerance of risks may depend on risk comparisons, the need to develop principles for risk comparisons is clear. Such principles are developed in Section 6.4.2 of the main report (UKAEA 1989) and may be summarised as follows:

- Different types of risk should be evaluated separately.
- Voluntary components should be excluded.
- Qualitative differences between the risks should be identified and, if possible, allowed for.

In addition, various problems encountered are considered when comparing risks (Sections 6.4 to 6.6).

It is concluded (Section 6.7) that "High consequence aversion" could in principle be applied to early deaths but no satisfactory theoretical way to establish a numerical representation (eg by CCDF shape) has been found.* Delayed fatalities, such as cancer deaths, should in any case be treated simply in terms of the expected total number of deaths, since these will be spread over a long period of time and will be indistinguishable, in statistical terms, from cancer deaths from other causes.

Lack of direct comparability of the hazards, and qualitative differences between the types of risk, cause significant difficulties in evaluating competing technologies eg between nuclear and conventional means of power generation.

Reference levels of societal risk of cancer from nuclear accidents may be derived either from the existing cancer risk or by comparison with background radiation; comparison of the risk of early death from nuclear accidents with the pre-existing risk from other accidental causes is possible but some care is required. However, it is concluded that it is not possible to construct from the data a CCDF which provides a useful reference level for risk evaluation; it may be helpful to bring the consideration of natural events into the analysis.

----

* *Note that this does not conflict with the acceptance of CCDF as a measure of risk.*

It is concluded that injury and birth defect risks from nuclear accidents may be controlled as part of a strategy for controlling the risk of early and delayed death respectively. However, on the basis of historical data, no useful reference levels have been found for evacuation and non-personal risks.

3.3 Existing Risk Targets

Risks from nuclear and other installations have been managed by regulation for a number of years. The targets developed in different countries are reviewed in the main report (UKAEA 1989 Chapter 7), particularly with regard to quantitative approaches.

In the EEC, following the "Seveso Directive" (EEC 1982), quantification of risk from non-nuclear as well as nuclear activities has been used as a guide to assessing acceptability and is recognised as useful in this context. This has been incorporated, for example, in the UK CIMAH (Control of Industrial Major Hazards) regulations. However there is a lack of agreement between national bodies in different countries as to the definition of an acceptable risk. In practice it is thought that quantitative assessment is generally regarded as an aid to decision making and a supplement to professional judgement.

Elsewhere, the situation for the non-nuclear risk targets is less well developed. The implementation of existing US requirements has been speeded up following the accident at Bhopal, and the chemical industry there has been involved in developing hazard evaluation and risk assessment procedures.

In the nuclear industry, techniques for quantifying risk have been developed over many years and development of criteria have progressed in line with the techniques available. The UK NII published their Safety Assessment Principles in 1979. However Layfield, in his report on the Sizewell Public Inquiry in 1987, called for the HSE to publish a consultative document as a first step in formulating guidance on tolerable levels of risk. The HSE document produced as a result, proposes an approach to risk targets with upper limits of tolerable risk and lower limits representing broadly acceptable levels below which "it would not be reasonable to insist on expensive further improvements to standards". This approach is adopted for individual risk. For societal risk, HSE seek to define the limit of tolerability in terms of comparison with other risks. The tolerability level however, is expressed in terms of radiation dose at a specified distance, which does not give a true measure of collective risk. Also the HSE say it is for the public, not the regulatory authority, to balance risks and benefits.

The WGRSPMA review concludes (UKAEA 1989 Section 7.6) that there is general unease about the application of the ALARP principle, and that clear guidance is needed on the precise conditions to be satisfied. Also the process by which the decision is reached should be open and consistent. There is general agreement on a level of tolerable individual risk to the public of death, of 10^{-6} per year. There is however a lack of agreement as to whether accidents causing multiple deaths should be treated proportionately more restrictively or not. Finally, there is general support for the use of probabilistic methods for safety assessment estimation, but less agreement on whether '*risk*' is the appropriate form for safety targets. It is recognised that economic arguments have to be taken into account in making decisions on acceptability, but there is no consensus on how this should be done. The use of optimisation, as defined by ICRP, does not offer any useful insights when applied to accidental releases, particularly major accidental releases.

4 RISK MANAGEMENT

4.1 Decision Theory and Decisions in Practice

Risk Management is concerned with taking decisions about risk tolerability. To apply a quantitative approach in this area it is necessary to consider both the decision process itself and the way in which quantitative targets can be set.

Modern decision making theory is a well established technique but it is little used in risk management. The reasons for this are discussed (UKAEA 1989, Section 8.2) and include:

- The theoretical impossibility of constructing an overall societal risk performance from the preferences of individuals.
- The difficulty of applying rules to derive subjective probabilities and utilities† in a coherent way.
- The aversion of society to assigning numerical values to utilities† of, for example, a human life.

Despite these difficulties, decision theory in the form of cost-benefit analysis is found useful and is applied where appropriate. Appendix 1 of UKAEA 1989 contains a proposed method for extending cost-benefit analysis to accidents.

In practice, decisions are taken by a complex process involving a number of bodies. In the UK, Public Inquiries have formed the focus for this activity in recent years. The following conclusions are drawn (UKAEA 1989, Section 8.3) on the basis of this limited experience:

(i) The importance of societal and individual risks in specific decisions has not generally been clearly presented in evidence, although the relevant data have been made available. Treatment of these issues has therefore been variable.

(ii) Despite this, Inspectors have found societal risks to be an important factor. The absence of any yardstick for societal risk comparisons has led to difficulties here, whereas individual risks have been easier to interpret in comparison with other risks. Individual risk comparisons have therefore been of greater influence.

(iii) Inspectors have not shown any indication of aversion to high consequence accidents.

(iv) There has been a degree of consensus that public opinion, and therefore risk perception, is an important factor in planning decisions.

4.2 Risk Targets

A number of risk targets are reviewed (UKAEA 1989, Section 8.4). The majority of those reviewed can be interpreted in terms of a "banded target" structure which consists of:

---

† *"utility" is a term used in decision theory to indicate some general measure of benefit.*

(i) An upper level of risk beyond which risks are intolerable.

(ii) A lower level of risk which is small and in some sense acceptable.

(iii) A range in between where risks should be minimised as far as reasonably practicable, based on judgement or a quantitative approach such as CBA.

This structure is recommended. Formal decision theory may be used in this context but should not be mandatory.

When using risk estimates in assessing compliance with quantitative safety targets, due account should be taken of the uncertainties in the estimates.

4.3 Societal Risk Management

Technological risk is tolerated by society because the benefits which result from the technology outweigh the risks involved. This is the fundamental social safety goal. However, whilst this concept forms the basis of decision theory and formal decision making techniques, these techniques are very difficult to apply in practice.

Historically, therefore, an alternative approach has been taken, which is to argue that there are levels of risks which are of no concern, and therefore their imposition does not require detailed comparison with the benefits. "Of no concern" may mean either that the consequences are small or the frequency is small. Difficulties are found, however, in identifying a consensus on these levels.

The WGRSPMA report (UKAEA 1989) is largely devoted to considering the technical basis for risk management. However, in practice, consideration must be given to the availability of adequate techniques for the analysis and also to existing practice. The available techniques are based on PSA which has been reviewed (see 2.5 above). With regard to existing practice, many targets exist and few are couched in terms of risk per se. There is a degree of uniformity between the differing existing targets, although care is necessary since the underlying assumptions may differ.

Targets are not set in terms of risk for two reasons: risk is a difficult concept to use in decision-making and secondly targets more specifically related to engineering quantities are more useful to designers and operators. However, an evolutionary approach is proposed to move towards risk-based targets by developing these from existing targets. An example of this is that proposed by the HSE in their recent consultative document. An alternative would be to use a CCDF of individual dose as the primary target. However, no specific approach is recommended.

A more controversial area is that of de minimis <u>exposure</u> or <u>doses</u>. It can be asked whether it is appropriate to base risk management on the hypothetical consequences of small radiation doses. The report recommends on balance that risk managers should consider the use of a '*de minimis*' dose level so that their policy focuses better on the true social impact of accidents.

It is clear therefore that risk management problems cannot be resolved entirely by scientific and technical study. There is an important societal dimension which the scientific and technical community must recognise. A symposium held at the University of East Anglia (UEA, 1988) considered this and the main points made are briefly mentioned below.

The safety of commercial nuclear power plants in England is a complex matter, which does, indeed, confuse the public. The responsibility for their safe design and operation belongs to the CEGB. They must, however, present a safety case to the HSE (in the form of the NII) which is responsible for licensing plants and for ensuring that they are operated within the licence. CEGB applications to build a reactor may also be subject to a Public Inquiry. Such an inquiry may consider safety issues in addition to planning and economic matters. The Inspector at a Public Inquiry can only make a recommendation: the final decision is made by the Secretary of State for Energy. He, and the Secretaries of State for the Environment and Employment (the minister responsible for the HSE), are accountable to Parliament and hence the public.

It is important that the public should have confidence in these arrangements, and the significance of this for the technical bodies involved is that their expertise and credibility should be accepted so far as possible. Unfortunately this is not necessarily the case at present. The reasons for this, and the ways in which this situation can be remedied, are beyond the scope of the report, but it is recommended that further efforts be made in this direction.

One aspect of this is the desirability of making the dialogue between the plant operators and the safety/licensing authorities more open and accessible to the public. There are many things which can contribute to this. In the technical area they concern mainly the presentation of information generated during the estimation of risk. This includes not only the final estimates of risk in some form, but also the associated explicit and implicit uncertainties.

PSAs are large, complex and hard to understand. But the most important information they contain is normally fairly simple: the dominant initiating event; the system reliability or physical process to which some measure of risk is most sensitive; the important judgements which have been made to produce a result; and so on. This information should be made more easily available - to the safety analyst, as well as the risk manager and the public. The underlying details should also be as open as possible to the scrutiny of those who care to examine them.

Finally, the technical community should try to explain its need for quantitative targets as an engineering necessity. These targets may not be wholly acceptable to the public, so the underlying reasoning should be explained as carefully as possible. And if rather arbitrary assumptions are necessary on technical grounds or simply to account for some aspect of public concern, then this too should be explained.

5 CONCLUSIONS

The main conclusions of the report are as follows:

(1) The regulation of individual risk does not necessarily ensure tolerable levels of societal risk in all circumstances, and so societal risk also needs to be controlled.

(2) Societal risk management should take into account at least five types of personal risk - early death, late death, serious injury, permanent or long-term evacuation, and serious birth defects - as well as non-personal harm to society, which is principally financial.

(3) It is not appropriate for primary quantitative risk targets to be set in terms of a single simple release parameter.

(4) The CCDF (or "F/N line") is recommended for evaluating societal risk, because it is the most accepted representation and the easiest to use and understand. This does not necessarily mean that targets should be set only using CCDF's, nor that CCDF's are uniquely preferred as a means of presenting risk data. The integrated measure of societal risk can give a useful coarse description, although it sometimes conceals important information.

(5) Existing risks in Great Britain and elsewhere have been estimated for both individual and societal risks. Accidents from man-made hazards become less likely as the number of casualties increases and the risk (viewed here as frequency x consequences) is roughly equally spread between large and small accidents. In contrast, large natural disasters involving many casualties dominate the risk from natural hazards.

(6) The estimates for existing risks show them to vary with time and the current trend is towards reducing risks.

(7) PSA can be used to estimate societal risks. However there are a number of areas where important uncertainties remain and therefore it is important that PSA results should contain quantitative estimates of the uncertainties associated with the important outputs (as far as reasonably practicable).

(8) Although '*high consequence aversion*' can be applied in principle to early deaths there is currently no satisfactory theoretical way to establish a numerical representation such as a CCDF shape. Cancer risks should in any case be treated simply in terms of the expected total number of deaths (subject to the associated uncertainty).

(9) Lack of direct comparability of the hazards concerned, and qualitative differences between the types of risk, make it difficult to compare nuclear societal risks with those from competing technologies. More attention needs to be paid to '*social risks*' from alternative sources of electricity, eg due to acid rain, carcinogen emission, and the greenhouse effect, from burning fossil fuels. This would offset the generally held view that only nuclear power is subject to large, catastrophic accidents with an impact on society as a whole.

(10) Reference levels of the societal risk of cancer from accidents may be derived either from the existing cancer risk or by comparison with background

radiation. However, for early death it is not possible from existing accident risk data to construct a CCDF which provides a useful reference level for risk evaluation.

(11) Injury and birth defect risks can be managed as part of a strategy for managing risks of early and delayed death respectively.

(12) No useful reference levels for evacuation and non-personal risks have been found from historical data.

(13) The use of banded targets is recommended. Risk management should be in the first instance based on levels of risk which can be shown to be low or high using reasoned arguments starting from reference levels of risk derived using appropriate principles.

(14) It is desirable for the institutional aspects of public safety to be as clear to the public as possible.

6 REFERENCES

AEA Technology, 1991, **Safety Policy and Corporate Safety Directives.**

BMA, 1987, **Living with Risk.** Wiley, Chicester.

Brown, M L, 1986, **Safety Aspects of EDRP.** UKAEA/BNFL Precognition at the EDRP Public Local Inquiry, D/P/8.

CEGB, 1982, **Design Safety Criteria for CEGB Nuclear Power Stations.** HS/R167/81 (Revised).

EEC, 1982, **Major Accident Hazards of Certain Industrial Activities.** Council Directive of 24 June 1982 (85/501/EEC). Official Journal of the European Community 5/8/85, L230/1 Vol 25.

HSE, 1988, **The Tolerability of Risk from Nuclear Power Stations.** HMSO, London.

I Chem E, 1985, **Nomenclature for Hazard and Risk Assessment in the Process Industries.**

Layfield, Sir Frank, 1987, **Sizewell 'B' Public Inquiry Report.** HMSO, London.

Royal Society, 1983, **Risk Assessment: A Study Group Report.**

UEA, 1988, **Risk Perception and Safety Targets for Major Accidents.** Report of a seminar held at the University of East Anglia, 16 October 1987. (Roberts, L E J, ed). Environmental Risk Assessment Unit Research Report Number 4, University of East Anglia.

UKAEA, 1987, **Code of Practice and Guidance Note: Radiological Guidelines for the Design and Operation of UKAEA Plant.** Safety and Reliability Directorate Report, SRD R 456.

UKAEA, 1989, **Social Risk: The Technical Basis for the Management of Risks to Society from Potential Major Accidents**, published as a companion volume to this document, 1989.

THE MANAGEMENT OF RISK TO SOCIETY
FROM POTENTIAL
ACCIDENTS

The main report of the UKAEA Working Group on the Risks to Society from Potential Major Accidents.

FOREWORD

In his report on the Sizewell 'B' Public Inquiry, Sir Frank Layfield expressed his concern that insufficient public information was available to "allow understanding of the basis for the regulation of nuclear safety". In particular, the lack of clear advice on acceptable levels of risk was considered to be significant. The Health and Safety Executive were charged with formulating and publishing such guidance, for both individual and social risk. The first stage of this has now been carried out. However, the Nuclear Industry in the UK has always maintained an active input into the decision making process, and AEA Technology has been concerned with risk targets, in one form or another, since they were first mooted in the late 1960s. Indeed, AEA Technology has published guidelines and a code of practice for Individual Risk Targets and has incorporated these into corporate safety policy and objectives. Progress was being made towards developing a social risk target when the accident at Chernobyl and the publication of Sir Frank Layfield's report changed the whole tenor of the basis for the discussions.

The treatment of 'social risk', as Sir Frank Layfield termed it, is concerned with the way in which potentially very large accidents are both guarded against in a regulatory sense, and can be tolerated by society in a sociological sense. The title chosen by the HSE for their advisory document uses the word 'tolerability' and not 'acceptability'. This is very important, as no accidents of such a magnitude as to cause social upheaval are acceptable, but they may be tolerable if they are both extremely rare and sufficient benefits to society can be identified from the activity. The AEA Working Group on Risks to Society in Potential Major Accidents was set up under my chairmanship to investigate the background issues involved in recommending social risk criteria; primarily to assist AEA Technology in making proposals for its own operations which reflected current thinking, but also to contribute to the debate stirred up by recent events. This document is the outcome of the Working Group's efforts. It does not recommend criteria, although it critically reviews existing ones. It is intended to provide a background to the diverse range of topics and technical disciplines which bear upon the subject. The Group was aided considerably in its work by consultants from outside the nuclear industry. In addition, a special seminar held at the University of East Anglia provided a most useful source of information from a range of relevant work in other fields.

The document is being published, and made widely available. We hope that readers will be stimulated by it, and will respond by sending comments back to us. The address to write to is given overleaf.

Finally, I must thank all the members of the Working Group who struggled hard and long with very difficult subject matter, to come up with what I believe is an important contribution to this topic.

Comments, enquiries or requests for further information should be addressed to:

Dr A R Garlick
SRD
AEA Technology
Wigshaw Lane
Culcheth
Cheshire
WA3 4NE

Acknowledgements

The United Kingdom Atomic Energy Authority in producing this book used figures from other sources. The UKAEA would like to thank the following organisations for their permission to do so: Elsevier, NRPB, IAEA, USNRC, Society of Risk Analysis, CAA

CONTENTS
(Main Report)

CHAPTER 1

INTRODUCTION

1.1 Background

Two events can be identified which are fundamentally significant for the production and contents of this report.

The first was the Chernobyl Nuclear Reactor accident in the Ukraine on 26 April 1986. This is the most serious nuclear accident experienced and it had a major impact not only on USSR society, but also on most other European countries and (in a different way) on those with any kind of nuclear power programme. These impacts were not entirely the same as those considered by risk analysts in the context of so-called societal risk. Most concern had been directed towards the health effects of radiological accidents, principally early death due to acute radiation sickness and delayed death due to the eventual induction of cancers. Because the number of delayed fatalities would be very much larger than those dying within a year or so of any accident, it was this effect which had appeared to be the most important. This was indeed the case for Chernobyl. The number of early deaths was 31, all of them on the site of the reactor, whereas the various estimates of the eventual number of cancer deaths run to several thousands throughout Europe. However, these large numbers are only one aspect of the impact of Chernobyl on Soviet and European societies. The others include: the forced evacuation, in some cases permanently, of large numbers of people; the massive decontamination effort required; the necessity to interdict foodstuffs not only in the country in which the accident happened, but also in countries thousands of miles away; and finally the enormous cost, running to several billion pounds, of making the Chernobyl site safe, replacing lost production both of electricity and foodstuffs, decontamination, resettling evacuees and loss of farmland. It was therefore appropriate for risk analysts, particularly in the nuclear industry, to reappraise their concerns and methods to make sure that the risk posed by potential major accidents in large industrial plants was suitably represented in terms of its impact on society.

The second event was the publication in January 1987 of Sir Frank Layfield's report on the Public Inquiry which had been held to make a recommendation on whether to grant permission to the CEGB to build a PWR at Sizewell. Layfield (1987) reached many conclusions, and made many recommendations which are relevant to the way decisions are reached in the UK about whether hazardous plants are 'tolerably safe.' In the area of risk tolerability he accepted the levels of individual risk to members of the public which the CEGB had used in deriving their Design Safety Criteria as tolerable "providing that there is expected to be economic benefit sufficient to justify the risks incurred." However, he was unable to reach any conclusion with regard to what he called social risk. He considered the field to be beset by serious problems which needed urgent elucidation. Of particular interest to risk analysts was his use of social risk expressed in terms of single numbers, rather than the frequency against consequence graphs which had become the norm for professional purposes. Layfield also expressed strong views about the obscurity of the decision-making machinery and the lack of guidance from Parliament on what constitutes a "tolerable risk", and how the legal requirement to make risks as low as reasonably practical (ALARP) should be implemented in practice. In this connection he recommended that the Health and Safety Executive (HSE) should formulate and publish guidelines on the tolerable levels of individual and social risk to workers and the public from nuclear power stations. Finally he urged greater public accessibility of the safety case for installations such as the Sizewell PWR and its use in licensing.

It was therefore timely for all bodies responsible for the safety of industrial hazards with the potential for large accidents to re-examine their policy, particularly with regard to the impact of such accidents on society as a whole. This responsibility is held by the plant designers and operators, but in the UK nuclear plant licensing is carried out by the Nuclear Installations Inspectorate (NII) of the HSE.

AEA Technology is responsible for the safety of operations on its own sites. It also carried out research into the safety of nuclear plant on behalf of the nuclear industry as a whole and so as to be able to advise the UK Government on such matters. Within AEA Technology, SRD, (AEA's safety and reliability consultancy business) acts both as a source of independent advice on the safety of operations and as a centre for safety research, development and consultancy to industry at large. SRD has a long-standing interest in the problems of risk assessment and risk acceptability, going back some 25 years to when F R Farmer, FRS, then Director of SRD, published a safety target closely related to societal risk. This target remains one of the few societal safety targets which have been seriously applied.

AEA Technology's policy on risk management has evolved over many years and is currently embodied in a Corporate Policy and Safety Directives document (AEA Technology, 1991), the basis of which on radiological risk matters is more fully explained in the associated Code of Practice and guidance notes on the control of radiological hazards (UKAEA, 1987). From 1990, AEA Technology has been subject to the same requirements for licensing of nuclear sites by HM Nuclear Installations Inspectorate as has the commercial nuclear industry in the UK. Thus current AEA Technology risk management policy and guidelines also reflect the current UK licensing requirements for nuclear plant generally.

The Working Group on the Risk to Society Potential Major Accidents (WGRSPMA) was set up in support of the development of AEA Technology corporate policy in the specific area of societal risk. It carried out the bulk of this work during 1988 and 1989. Its chairman and secretariat were SRD based but its membership included safety experts from throughout AEA Technology and from other bodies.

The Working Group decided that a review of technical issues related to societal risk would be useful both within the nuclear industry and more generally. This is that review; its objectives are to:

- provide the essential background for the interested non-expert, particularly via the references
- outline the major problems in the field
- report on work carried out by the Group to throw light on some of these areas, and
- make general recommendations on matters such as definitions, methods for estimating and representing risk, principles for risk acceptability and means of making decisions.

Because of its background as part of AEA Technology's policy formulation process, this document, while considering risk issues in general, makes particular reference to their relevance to the nuclear industry. But it is not an objective of this document to formulate specific quantitative safety goals for AEA Technology or any other body.

Since completion of this study, AEA Technology has become the trading name of the United Kingdom Atomic Energy Authority (UKAEA), and the names UKAEA and AEA Technology are both used in this document. The UK electricity supply industry has been privatised, and the nuclear operations of the former CEGB are now carried out by Nuclear Electric; CEGB is more often referred to in this document, particularly in the context of the case made and extensively analysed at the Sizewell 'B' inquiry.

1.2 Scope of the Report

To clarify the structure of this report we begin with a few definitions taken from the Royal Society Study Group report (Royal Society, 1983). This is an earlier review of the subject prior to Chernobyl and Layfield.

> The general term used to describe the study of decisions subject to uncertain consequences is RISK ASSESSMENT. It is conveniently sub-divided into RISK ESTIMATION and RISK EVALUATION. The former includes:
>
> (a) the identification of the outcomes;
>
> (b) the estimation of the magnitude of the associated consequences of these outcomes; and
>
> (c) the estimation of the probabilities of these outcomes.
>
> RISK EVALUATION is the complex process of determining the significance or value of the identified hazards and estimated risks to those concerned with or affected by the decision. It therefore includes the study of risk perception and the trade off between perceived risks and perceived benefits. RISK MANAGEMENT is the making of decisions concerning risks, and flows from risk estimation and risk evaluation.

Thus, as implied by its title, this report is concerned with the technical (or professional, or scientific) basis for taking decisions about risks. However, this is not a topic which can be resolved on a purely scientific basis. In fact there are three basic problems which impose constraints on the risk manager's policy.

The first of these is the capability of the risk estimation technique used. Decisions have to be taken in the light of available information, and this is inevitably incomplete and uncertain. In fact it is the development of probabilistic techniques which has made it possible to consider safety goals formulated in terms of risk, by comparing different risks on a single scale or by trading off risks with benefits.

The second, and related, problem is that of ensuring that any policy is capable of implementation. That is, the designers and operators of hazardous plant must have the engineering tools available to make a safety case which can be assessed against the safety goals.

The third problem is caused by the fundamentally political nature of risk acceptability. Thus, although the responsibility for plant safety is held by identified individuals, they are accountable to the general public via a series of bodies involving Parliament and ministers. The safety case therefore has to be accessible and acceptable to the various groups involved, and this introduces matters such as the perception of risk by the public. For this report it is the total aversion of the public to large nuclear accidents which is of particular importance. The boundary between scientific and political matters is neither clearcut nor static and the Working Group has neither the expertise nor the remit to examine in detail the political constraints imposed on risk managers. However, it is appropriate in a technical document to report and comment on those aspects of the public perception of risk and risk assessment which have been scientifically established and this we shall do where possible.

The report begins by clarifying the concept of risk and defining it suitably in Chapter 2. In particular the two aspects of individual and societal risk are separated. (Note. Layfield adopts the term social risk. It is not clear why; the word societal is in widespread use in this field and we shall continue to use it along with Layfield's neologism.) It is inevitable that it is not societal risk alone that is discussed in the remainder of the report. The first reason for this is that the field of individual risk tolerability is relatively well developed and can thus provide pointers for societal risk. The second is that, to a large extent, societal risk is the aggregation of individual risks. This is described in more detail in Chapter 3 which examines the various types of risk, how they can be measured, and how they can be represented. In doing this it also provides an overview of how quantitative safety goals can be set. Chapters 2 and 3 thus form the background against which the other risk management activities can be examined.

Risk estimation and risk estimates form the subjects of Chapters 4 and 5. Chapter 4 presents direct estimates from statistical data of the levels of individual and societal risk which currently exist. They form useful material for the risk evaluation stage which will be heavily dependent on risk comparisons. Chapter 5 discusses indirect risk estimation. This is distinguished from direct risk estimation by the need to provide a model of the hazard which enables the risk it poses to be deduced from the behaviour of its constituents. It is the technique which has to be applied to major hazards because of the absence of data for direct estimation. The discussion in Chapter 5 is aimed at examining the suitability of the technique for assessing whether a given risk is tolerable. It is thus intended as a review of risk estimation for risk evaluators, not risk estimators.

Chapter 6 is concerned with risk evaluation. It covers the principles of risk comparability and risk acceptability or tolerance. (Note. Layfield rightly argues that a risk should not be described as acceptable just because it is tolerable. We support this, though for reasons of style and authors' preferences (see next section) the term 'acceptable' has not been eliminated from this report.) It is here that risk perception issues enter; apart from pointing out the need for risk managers to bear public opinion in mind, the chapter examines ways in which useful reference levels of risk can be defined.

Chapters 7 and 8 discuss risk management proper. Chapter 7 reviews the safety policy and quantitative targets which have been proposed by various regulatory bodies. It concentrates on the UK but also covers the EEC and the US. The details of this provide further useful leads for determining the most appropriate approach. Chapter 8 is concerned with decision making per se. It covers a spectrum of material ranging from the mathematical theories which prescribe idealised decision making techniques, through to an examination of how decisions have been taken in practice by inspectors at public inquiries, for example. Some additional material on cost benefit analysis is included in Appendix 1.

Chapter 9 provides an overview of the complete problem, drawing on the material from the preceding chapters, and pulls together the threads to propose a broad approach to the formulation of risk management policy taking practical considerations such as existing procedures into account. The conclusions and recommendations are recapitulated in Chapter 10.

A glossary and list of acronyms is provided in Appendix 2.

1.3 WGRSPMA Membership and Authorship of the Report

The Working Group on the Risks to Society from Potential Major Accidents had a membership drawn from all UKAEA sites, together with three experts from outside the UKAEA. The SRD members were: Dr M R Hayns (Chairman), Dr A A Debenham, Mr A R Taig and Dr A R Garlick (Secretary). Other UKAEA members were: Dr R Bullough FRS (Harwell), Dr F Briscoe (Culham), Dr C F Clement (Harwell), Mr P L Holden (Corporate Headquarters), Mr A Neal (Winfrith), Mr P A H Saunders (Harwell), Mr R Shallcross (Dounreay) and Dr D Wilkie (Windscale). External members of the Working Group were: Dr J C Chicken (JC Consultancy), Ms D P Fernandes-Russell and Prof L E J Roberts FRS (both from the Environmental Risk Assessment Unit, the University of East Anglia). In addition the Group was assisted by Dr I Cook (Culham) and, in a reviewing capacity, Mr E V Gilby (Gilby Associates).

As is usual in documents of this sort, each chapter was initially produced under the lead authorship of one of the Group members. It is inevitable, and indeed desirable in view of the well known properties of committee productions, that these chapters will particularly reflect the views and perspectives of their authors. However, the whole report has been extensively discussed by the Working Group and reflects the joint views of its members to the greatest extent possible with such a project.

1.4 References

AEA Technology, 1991, **Safety Policy and Corporate Safety Directives.**

HSE, 1988, **The Tolerability of Risk from Nuclear Power Stations.** HMSO, London.

Layfield, F, 1987, **Sizewell 'B' Public Inquiry Report.** HMSO, London.

Royal Society, 1983, **Risk Assessment: A Study Group Report.**

UKAEA, 1987, **Code of Practice and Guidance Note: Radiological Guidelines for the Design and Operation of UKAEA Plant.** Safety and Reliability Directorate Report, SRD R 456.

CHAPTER 2

THE NATURE OF RISK

2.1 Introduction

This report is concerned with risks to society from potential major accidents. Because of the origins and needs of the work, it is aimed primarily at the nuclear power industry. In the conditions prevailing after years of debate over the tolerability of nuclear power, and brought to a head by Chernobyl, it is important, perhaps crucial, to concentrate on those accidents with the potential to have a societal, rather than just an individual, impact. The position of the border line between accidents which have, or do not have, societal impact is open to debate, but it is clear that at some point the synergistic result of consequences which are both large and diverse is a total effect which is larger than the sum of its individual parts. It is with such accidents, and the ways in which 'acceptance' or 'tolerability' levels may be established, that this report is concerned.

Having studied the subject, we are under no illusions about the tractability of the problem. The subject of risk in a high technology society opens up many areas of study beyond the technical competence of the authors of this report and the committee that edited it. We are, in the main, nuclear technologists and our concern is the safe design and operation of nuclear reactors and their associated plant in order to realise the immense energy resource contained in nuclear fission. In trying to establish engineering targets and goals for operational safety, it is necessary to provide a yardstick to aim at. It is trying to decide what this yardstick should be in terms of the risks posed by the plant to the public that is difficult. This involves areas of science and sociology unfamiliar to many of us.

In order to set the scene, this chapter sets out the problems which we believe need to be addressed and hence provides an overview of the specific problems to be addressed by each chapter. It begins with a discussion of the historical, cultural and economic background to the concept of 'risk' in section 2.2. We then introduce the formal definitions of risk in section 2.3 and extend them to individual and social risk in section 2.4. Finally, section 2.5 sets out the thinking we believe ought to be applied in order to resolve the gap between the needs of the scientists and engineers using the technology, and the public who are exposed to the hazards from it.

2.2 The Concept of Risk

2.2.1 A brief history

'Risk' has been at least acknowledged, if not understood, since the earliest times for which historical records are available.

A comprehensive review of the development of risk analysis and risk management by Covello and Mumpower (1985) paints a picture of the concept of risk evolving from exposure to misfortune and the vagaries of our natural environment to exposure to industrial hazards. Thus, the concept of risk is deeply embedded in our cultural heritage; it may even be part of our race memory from pre-history. That much of the early evaluation of risk as a concept was closely linked with the development of religious thought, especially the probability of the after-life, should give the modern student of the subject a strong forewarning that these are deep and difficult waters.

The concepts of insurance and home commercial risk management have a history almost as long as the concept of risk itself, but the real milestone came with the development of probability theory by Pascal in 1657. This seems to have initiated a flurry of activity culminating in the first quantitative assessment of risks to health which would be recognised by modern practitioners - LaPlace's analysis of the influence of smallpox vaccinations on the probability of death in 1792.

An appreciation of the risk to the environment as a variant of the concept of risk to people began to emerge only later. (We leave aside many early assessments of risk to farmers from flooding, pestilence and so on.) Indeed, the connection between risks to man and risks to the environment, and hence the concept of the symbiosis between man and his environment, came later still. This is described in Lord Ashby's (1978) seminal book on environmental risk. The attainment of a society with time and wealth to appreciate the natural environment seems to have been a prerequisite for such a development. Perhaps the most eloquent example of such thinking is to be found in "Rousseau's Nightingale," as cited by Ashby. Rousseau asks what lengths we should go to to have the pleasure of hearing the nightingale sing.

As we come closer to the present time, and particularly the last decades, the amount of literature expands enormously and the historical perspective is lost. However, the concept of 'risk' does seem to have continued to evolve, and certainly the common usage of the word is in flux. The emergence of the environmental movement has served to focus on the relationship between man and his environment to the point where it is no longer simply the impact of an activity on people that is of concern, but rather the wider feeling that we must be cognisant of the need to protect our total environment that is of prime importance. These are not altruistic or sentimental feelings. The clear interdependence of all life on earth is now widely accepted, though few would go quite as far as Lovelock with the concept of Gaia. This movement has Rachel Carson's 'Silent Spring' as one of its influential milestones. Whether or not we agree with the technical details of works such as this, we must accept that they have significantly affected the intellectual climate in which we now work, and we should attempt to express our goals in terms which properly match the public's perception of these matters.

This brings the development of the concept of risk to the present. We have finally introduced the idea of <u>perception</u> of risk. Risk cannot be felt (although fear can) and it is essentially unmeasurable. It can, however, be calculated, albeit imperfectly, and this leads to further problems. Take, for example, the question of life expectancy. This has improved enormously this century. Increases of about 20 years have been achieved in Western Europe and North America (BMA, 1987). Despite this, reductions in the frequency of catastrophic events and continuing assurances that the health of the population is getting better, people constantly indicate through polls and other sampling techniques that they believe life is getting riskier.

Furthermore, continuing scientific investigations are bringing new and previously unknown risks to the attention of the public, creating an impression of an environment becoming increasingly hazardous for its inhabitants. Research into the factors controlling people's perception of risk has indicated that the primary attributes for public concern are not mortality or morbidity rates, which seem remote, but characteristics such as: the potential for catastrophe; lack of familiarity and understanding; the involuntary nature of risks; scientific uncertainty; lack of personal control; risks to future generations; doubtful benefits; inequitable distribution of risks and benefits; and potentially irreversible effects. When coupled to the dread of nuclear matters associated with weapons and the 'mystery' of radiation in general, these factors give some insights as to why the generation of electricity by nuclear fission is bearing the brunt of the debate concerning risks to society from the activities of advanced technological industries.

2.2.2 Economic and cultural factors

The evolution of 'risk' as a component of our culture has been presented above from the rather parochial view of western cultural development. Any consideration of the acceptability (or tolerability) of risk is therefore dependent on the cultural values existing in a society. Of particular concern is the relationship between purely economic factors (essentially a matter of insurance) and other aspects of society. Thus, the value attributed to human life has important ramifications when judgements are made as to the costs which would be considered appropriate to improve the safety of plant, and hence save lives. This question is addressed later in terms of the applicability of cost benefit analysis to accidents large enough to have social implications.

Different nations and cultures might have differing approaches to the issues of the cost of life and the balance between the desire to develop technologies with the potential for great benefit and the need to make plant safe. We shall not go into this, but simply note that we are addressing this issue from the perhaps parochial needs of an industry operating in the conditions of the UK.

The central issue concerning the operation of hazardous plant is the requirement for environmentally and socially aware countries to come to terms with the technologies which make its way of life possible.

2.3 Definitions of Risk

Risk is defined by the Oxford English Dictionary as "hazard, danger; exposure to mischance or peril". This definition, however, does not reflect the components of risk, nor does it give any indication that it is ubiquitous. There are several definitions available from many technical sources, though a recent OECD (1986) report concluded that no compatible set of definitions in the risk field existed in member states. One authoritative definition was given by the Royal Society Study Group (Royal Society, 1983), and is much quoted:

> For the purposes of this report the Study Group views RISK as the probability that a particular adverse event occurs during a stated period of time, or results from a particular challenge.
>
> An ADVERSE EVENT is an occurrence that produces harm.
>
> With RISK defined as above, HAZARD is seen as the situation that in particular circumstances could lead to harm, where HARM is the loss to a human being (or to a human population) consequent on damage and DAMAGE is the loss of inherent quality suffered by an entity (physical or biological).
>
> BENEFIT is the gain to a human population.
>
> DETRIMENT is a numerical measure of the expected harm or loss associated with an adverse event.

This definition, or group of definitions, is rather general, though comprehensive. A set of definitions has been developed by the Institution of Chemical Engineers (I Chem E, 1985) suitable for use in the chemical process industry:

> **HAZARD; a physical situation with a potential for human injury, damage to property, damage to the environment or some combination of these.**

> RISK; the likelihood of a specified undesired event occurring within a specified period or in specified circumstances. It may be either a frequency (the number of specified events occurring in unit time) or a probability (the probability of a specified event following a prior event), depending on the circumstances.

These two sets of definitions are typical of a number which have been produced, but which are of the same type. The difference between them shows the desirability of using terms which are most suitable in a particular context. For example, the British Medical Association in their book 'Living with Risk' (BMA, 1987) choose the Royal Society Study Group's definition. In our case we need a rather more technical description and therefore adopt the I Chem E definitions. We use the definition of 'hazard' as above, but change the definition of 'risk' slightly to emphasise that the risk due to a plant is composed of the possibility of many different adverse events. We also remove a slight inconsistency in the definition of the numerical quantities:

> **RISK; the likelihood of specified undesired events occurring within a specified period or in specified circumstances arising from the realisation of a specified hazard. It may be expressed as either a frequency (the expected number of specified events occurring in unit time) or a probability (the probability of a specified event following a prior event), depending on the circumstances.**

This definition shows that we are concerned with a range of potential accidents, causing different types of harm and to differing extents. We shall narrow this down somewhat when we introduce individual and societal risk in section 2.4.

Although both sources above mention probabilities in their definitions, neither defines what is meant. In fact the nature of probability - which defines the nature of risk - has been a topic of debate by mathematicians and philosophers ever since its concepts were first applied. We do not wish to go into this in any detail; the definitions are simply saying that we elect to represent risk quantitatively using the algebra of probability. This is discussed further in Chapter 5 which is concerned with estimating these probabilities, and in Chapter 6 where the rationale for comparing risks is dependent, in part, on the nature of the probabilities in the risks to be compared.

The three types of probability which occur when estimating the risk posed by nuclear plant are:

- those obtained directly from observations (such as the statistics of road accidents),
- those obtained by logical deduction (with such techniques as the fault or event trees of probabilistic safety assessment),
- those expressing degrees of belief (such as arise from the techniques used to extract expert opinion).

The nature of these three types of probability, as well as the difference between frequency and probability, need to be borne in mind whenever discussions of risk, and particularly its quantification, are undertaken.

2.4 Social Risk

2.4.1 Individual and social risk

Risk, as defined previously, is a combination of two quantities - the probability that an unwanted event will occur and the consequences of that event. The expression of the nature of the consequences is important as it defines the kind of risk we are concerned with.

A measure of risk which is useful to designers, operators and managers of plant is individual risk, and this is often used for risk management. Take, for example, a target of 10^{-7} per year for the risk of death to a member of the public from a particular accident leading to release of radioactivity. (We note that to have a societal impact there would need to be a major accident, with offsite releases of radioactivity and we leave aside the questions of risks arising from permitted operational releases and the possible societal impact of low level releases on a local community.) Such a criterion is well established. Thus, for example, if an accident was determined to have the potential to deliver a dose of 10 mSv to a member of the public, this dose would correspond to a probability of premature death of about 10^{-4} (using the dose-risk relationship recommended by the ICRP (1977); note, though, that higher values are implied by new guidance given by the NRPB (1987)). Therefore, in order to meet the risk target, it would be necessary to show that the probability of the accident occurring and delivering this dose to the individual was less than 10^{-3} per year (10^{-3}x10^{-4}=10^{-7}). This form of criterion is easy to use since it gives direct advice to the designer or operator on what levels of reliability are called for to achieve the set targets. This also applies to individual risk targets which are used to control the doses to the workforce. The above example is taken from the safety precognition to the EDRP public inquiry (Brown, 1986).

The use of individual risk could be criticised in that it might miss important factors when the total effect of a release is taken into account. Further, for safeguarding the workforce, the collective dose (see Appendix 2 for a description of the various 'dose' terms used in radiological protection) is also an important factor: limiting this prevents the use of large numbers of people in a high radiation environment, each receiving an 'acceptable' dose, but which in toto would give unacceptably high consequences. Consider the example quoted above. A dose of 10 mSv, with a probability of 10^{-3} per year does indeed give an overall risk to an individual receiving that dose, of 10^{-7} per year. However, if the number of people receiving that dose was 10^6, say, then the individual risk target would 'let through' accidents capable of killing 100 people.

An additional target is needed to deal with accidents where the size of the consequences becomes large enough that the maximum individual risk does not fully represent the societal impact of the accident.

In addition to the straightforward problem of recognising the aversion to large consequence accidents, and hence guarding against them, we have to recall that the 'consequence' aspect of risk covers a range of quantities. These are described in detail in Chapter 3. For example, a situation might arise in which individuals are protected against the risk of death but society finds the cost of frequent evacuation after small accidents unacceptable. Indeed, this illustrates a general principle that different types of risk are not directly related and may even be traded off against each other. For example, failure to carry out a radioactive waste disposal programme may reduce risk to the public, but at the expense of increasing worker risk.

The argument can be reversed. Technological risks are imposed because they are thought to be outweighed by the resulting benefits to society. It could then be claimed that the fundamental risk assessment be carried out on a societal basis. The function of individual risk targets in this picture would be to ensure that the risk burden did not fall unfairly on a small number of individuals in the society. The problems of comparing risks with benefits are outlined in Chapter 8, and these explain in part why individual risk targets are better developed than those for society. The present discussion indicates the source of some of the problems in technological risk acceptance in a democratic society. Those suffering the risk, which is imposed involuntarily, are not necessarily those gaining the benefits; indeed, they may not recognise that there is a benefit. Certainly, there is no agreed procedure for estimating and comparing risks and benefits.

This shows that individual and societal risk address fundamentally different concerns. Society as a whole consists of a collection of overlapping, and sometimes conflicting interests (national good, company profit, environmental amenity, individual safety ....) and the groups representing each interest take a different view of the risks and benefits of any development. Societal and individual risk guidelines are aimed at ensuring the correct approach to the first and last of the interests mentioned respectively. This also indicates that tolerating risks is a fundamentally political matter; we discuss this further in section 2.5.

A corollary of this argument is that the notion of consistency between individual and societal risk targets is not coherent. Such consistency arguments have sometimes been used to adduce societal risk targets from individual risk targets (for example, the Dutch criteria described in Chapter 7).

The way in which individual and/or societal risk targets are used depends on the risk management context. Using these concepts inappropriately has led to misunderstanding and potential conflict - a point we also discuss further in section 2.5.

With this background we turn to defining societal risk and begin with the definitions put forward by the I Chem E (1985):

> **INDIVIDUAL RISK; the frequency at which an individual may be expected to sustain a given level of harm from the realisation of specified hazards.**
>
> SOCIETAL RISK; the relationship between frequency and the number of people suffering from a specified level of harm in a given population from the realisation of specified hazards.

Although the latter definition is adequate for our purposes, it does not emphasise the many types of harm which may befall people, nor does it mention aspects which apply to society as a whole (see Chapter 3). Finally, it seems rather inconsistent to define a risk as a relationship. We therefore propose the following definition:

> **SOCIETAL RISK; the frequencies with which specified numbers of people in a given population, or the population as a whole, sustain a specified level of harm from the realisation of specified hazards.**

Since this definition is purely descriptive, it does not help in defining a quantitative 'social risk goal'. However, we feel it is useful to have a simple statement which attempts to embody the ideas described at some length in this chapter. It should be understood and interpreted in the context of the discussion and the danger of over-simplification in such a richly complex area should be kept in mind.

2.4.2 The purpose of social risk targets

We have seen that the purpose of societal risk targets is to protect society as a whole. The situation is slightly paradoxical in that although the nature of the topic is the global weighing of risks and benefits, the way in which the risks are quantified is determined to a large extent by the concerns of the individuals in society. Thus, societal risk is particularly associated with the study of large accidents with many casualties, and it is frequently taken for granted that 'risk aversion' should be incorporated. That is, society is more concerned with single accidents causing 100 deaths than with 100 accidents each with a single fatality. The reason for this is that such accidents have a larger impact on society, because of:

- media coverage,
- the greater uncertainty in the frequency of such accidents,
- the perception that institutional control of hazards has failed, and
- the greater disruption of family and community.

We shall refer to this as 'high consequence aversion', a term which better describes this effect than 'risk aversion'.

The idea of social impact is thus useful for examining other aspects of social risk. For example, the importance of the forced permanent evacuation of communities has been underlined by the Chernobyl accident, as has the cost to society of all the activities which are necessary following a major nuclear accident. All these types of risk are discussed further in Chapter 3. This idea also emphasises that this report is concerned with major accidents; thus one of the subjects that will not be addressed is that of the social (or collective) risk of workers at major hazard plant.

The general objectives of the report and the topics to be covered were set down in Chapter 1. The foregoing discussion indicates the main purpose of each Chapter in the review (chapter numbers in brackets):

- to identify those aspects of technological risk which have a significant social impact (3),
- to examine the extent to which these risks can be estimated numerically (5),
- to consider which of these risks can be meaningfully compared with other risks which exist in society (6).
- to estimate such existing risks (4),
- to consider the technical aspects of the tolerance of risk by society, its institutions and its individuals (6),
- to review existing risk management policy so far as seems relevant (7),
- to discuss how decisions are made in theory and in practice in implementing policy (8), and
- to draw conclusions and recommendations from these which relate to the formulation and implementation of social risk policy (9 and 10).

2.5 Risk Acceptance in the Decision Making Process

Formulating and implementing risk policy has both technical and political aspects. Broadly speaking, the formulation is political and the implementation is technical. There is no well-defined boundary. Indeed, there are instances where the overlapping interests of policy makers and technicians cause confusion, and others where there is no overlap at all and a gulf exists between the two communities. This is addressed in more detail by Hayns and Gittus (1987). For example, politicians can call for absolute safety - which of course cannot be implemented. Similarly, technicians may call for public risk acceptance to be based on 'rational' criteria such as the comparison of risks or the use of decision theoretic concepts (see, for example, Hayns and Unwin (1985)). As long as the public remain unimpressed by, and incomprehending of such rationality, its political usefulness is limited.

We therefore wish to stress the importance of analysing the political nature of risk management. For example, the organisation and credibility of the relevant institutions may be more significant than the precise nature of the technical criteria. However, as technicians, our main concern is with defining technically implementable risk management policies.

We have not ignored the problems of public perception and political reality entirely. They are, of course, also susceptible to scientific analysis and a symposium was held at the University of East Anglia under the chairmanship of Prof L E J Roberts to discuss these aspects with academic experts in diverse areas. A report on the discussion has been published (Roberts, 1988). The principal lesson from the meeting was that there is a need to establish the public credibility of the institutional provisions for public safety. This is not achieved by the use of quantitative assessments and numerical targets, which the public find inaccessible and unhelpful, but by establishing good practice and engendering confidence.

This discussion draws attention to the different technical needs of the various interested parties. Designers of plant require detailed numerical targets such as those in the CEGB Design Safety Guidelines discussed in Chapters 7 and 8. Post-accident emergency planners need a similar type of information. Regulators need to frame non-technically worded legislation and other safety goals in terms of technical procedures or structured technical judgements, a process illustrated by the Nuclear Installations Inspectorate's Safety Assessment Principles, also discussed in Chapters 7 and 8. Risk managers may require condensed forms of numerical information; for example, the overall risk targets in the CEGB Design Safety Criteria, or the single measure of social risk used by the Inspector at the Sizewell 'B' Public Inquiry. In general, the level of information depends on how far the manager is removed from the design/operation environment. The technical information required by the public may vary from none at all to detailed, quantified fault trees.

Because we are mainly concerned to give guidance to designers and operators, and because of the unresolved political and technical problems in the acceptance of risk, we shall need to define levels of risk which can be regarded neither as 'known to be tolerable' nor as 'ought to be tolerable'. Instead, we examine the approach taken and levels proposed by other authorities in order to identify an approach to social risk which is broadly consistent with what has been done previously. In doing this, we have critically examined the reasoning involved, and attempted to make clear what the assumptions are.

2.6 Conclusions

(1) We have provided formal definitions of risk in terms of probabilities, but the general concept of risk is complex with historical, cultural and economic aspects.

(2) The regulation of individual risk does not necessarily ensure tolerable levels of social risk in all circumstances. Therefore social risk also needs to be controlled.

(3) We recognise the needs of various parties - regulators, politicians and the public - but we are mainly concerned with the problem of providing technically based guidelines to the engineers and scientists who actually build and operate the plant. Thus we shall try to establish guidance which, when implemented, would result in broader risk acceptance criteria being met.

2.7 References

Ashby, E, 1978, Reconciling Man with the Environment. Oxford University Press.

BMA, 1987, Living with Risk. Wiley, Chichester.

Brown, M L, 1986, Safety Aspects of EDRP. UKAEA/BNFL Precognition at the EDRP Public Local Inquiry, D/P/8.

Covello, V T, and Mumpower, J, 1985, Risk Analysis and Risk Management: An Historical Perspective. Risk Analysis, 5(2), pp 103-120.

Hayns, M R, and Gittus, J H, 1987, Risk Assessment. Proc Roy Soc Edinburgh, 92B, pp 139-154.

Hayns, M R, and Unwin, S D, 1985, Rational Quantitative Safety Goals. ANS Winter Meeting, San Francisco.

I Chem E, 1985, Nomenclature for Hazard and Risk Assessment in the Process Industries.

ICRP, 1977, ICRP Publication 27. Annals of the ICRP, 1)4), p 1.

NRPB, 1987, Interim Guidance on the Implications of Recent Revisions of Risk Estimates and the ICRP 1987 Como Statement. NRPB-GS9, Chilton, Oxfordshire.

OECD, 1986, The Need for Common Definitions of Risk. OECD Report ENV/ECO/86.2.

Roberts, L E J, (Ed) 1988, Risk Perception and Safety Targets for Major Accidents. Report of a seminar held at the University of East Anglia, 16 October 1987. Research Report No 4, Environmental Risk Assessment Unit, University of East Anglia, Norwich.

Royal Society, 1983, Risk Assessment: A Study Group Report.

CHAPTER 3

CHARACTERISING AND REPRESENTING RISK

3.1 Introduction

In this chapter we describe a number of proposed ways of quantifying and presenting risk. This is particularly important for nuclear risks which are of a different nature to many common risks. If comparisons are to be made between risks, they must have some common measure.

The definitions and discussion in Chapter 2 show that although risks are defined in terms of the probabilities (or frequencies) of specified harmful events, the nature of the events is left open. The choice is restricted to some extent by considering only individual and societal risk, but it is evident that there are many types of harm which may be sustained. Correspondingly, these give rise to various types of risk, and in section 3.2 we outline the most important types, which are sub-divided into personal and non-personal, that is, purely social.

There are a number of aspects of risk which are of particular interest to the nuclear industry and two of these are discussed in section 3.3. The first is the complications caused for risk representation and comparison by the delayed effects of radiation. Secondly, quantitative safety goals do not necessarily have to be applied in terms of risk. They can apply to individual system reliabilities (see, for example, the CEGB Design Safety Guidelines for PWR Reactors (CEGB, 1982b)), to the frequency of releases (Farmer, 1967), or in many other ways (USNRC, 1981). The relationship between targets of this kind and risk is discussed in subsection 3.3.2.

Section 3.4 deals with the various ways in which it is possible to present risk (that is, the probabilities of various events). Here it is societal risk which is of interest since we wish to consider many different events simultaneously such as the frequency of accidents killing 1, 2, 3,....998, 999, 1000,.... people.

3.2 Types of Risk

We now need to specify the types of harm (or consequences) arising from possible undesired events. If these consequences can be quantified, and an estimate of the frequency is available (whether directly from observed data (Chapter 4) or indirectly from modelling a system (Chapter 5)), we have a measure of the risk of the event. In the past, most individual and societal risks have been expressed as the probability of immediate death or the number of early deaths. This consequence is manifestly inadequate to describe the risk of a major nuclear accident where very few, if any, people would die immediately. In addition, we must consider other consequences and we divide them into two types, personal and non-personal. The non-personal types refer to harm to society which is not incurred individually by each member. They are largely financial in nature.

3.2.1 Personal risks

Risks which directly affect individuals may be termed personal risks. For serious consequences to individuals, we introduce the idea of a 'risk profile' of a disaster to characterise the numbers of people who would be affected in different ways by it. That for Chernobyl is shown in Figure 3.1 and is characterised by the five headings: early death, late death, serious incapacity, forced permanent evacuation and serious birth defects. The first four of these represent risks to living individuals and the final one is a risk to future individuals. Serious incapacities include loss of limb, blindness and inability to work, but the boundary with minor injury might not be always easy to draw. Also the categories of late death and serious incapacity are not mutually exclusive, as appears to be the case for some victims of Bhopal whose lungs were damaged. Such cases should appear in both categories. For the delayed cancer death in the nuclear case the victim is not incapacitated during the latent period.

For each category, a societal risk can also be obtained in terms of the number of people affected, consistent with the definition in Chapter 2. We call the sums N_i where N_1 is the number of early deaths, N_2 the number of late deaths, N_3 the number of people seriously incapacitated, N_4 the number of people permanently evacuated and N_5 the number of serious birth defects. The sums and the risks they represent are quite different for different major accidents. For a major chemical (Bhopal) and a major nuclear (Chernobyl) accident the differences are illustrated in Table 3.1.

It would considerably ease the study of societal risk if a single measure of risk could be obtained by summing over the five types. In Chapter 6 we shall say why we do not consider it appropriate to compare the types, and thus we do not recommend carrying out such a sum. However, we continue this review by noting some attempts which have been made to do this. The simplest comparison could be made by constructing the sum:

$$N = \sum_{i=1}^{5} w_i N_i$$

where the wi are weighting factors to compare the risk with that of early death, $w_1 = 1$.

At one extreme, we could use just the total number of people affected and take wi = 1 for all i (but allow for double counting in categories 2 and 3). However, this does not allow for society's views on the seriousness of the consequences. When given the choice, people nearly always choose evacuation rather than risk death or incapacity. This is the basis of hurricane warnings on the coast of the US which save many lives.

Some proposed methods to choose the wi are now described.

There is a strong argument that the comparison between early and delayed death could be made on the basis of loss of life expectancy; this comparison has been advocated for the nuclear risk of delayed cancer deaths by Marshall et al (1983). As they emphasise, this method takes into account the possibility of intervening death from another cause, and also enables a comparison to be made with the risk of smoking. Loss of life expectancy is discussed further in subsection 3.3.1 and in Chapter 6 where a value of w_2 is about 1/3 is estimated on this basis. Alternatively the work of Kinchin (1978) and Levine (1981) uses a factor of 1/30. This illustrates the problems involved in evaluating such factors.

The quality adjusted life-year (QALY) is a concept first proposed in the US which has been developed in the UK by Professor Alan Williams and his colleagues at York University (Williams, 1985; Kind et al, 1982) as an analytic criterion for the assessment of potentially beneficial health care procedures or, with associated costs, as a measure of 'value for money' in health care. Each year of life is multiplied by a fraction expressing the impairment of the quality of life experienced by survivors. The practical difficulties in developing a set of quality adjustment fractions are formidable, and the concept has been criticised as presenting a political issue as a technical issue (Smith, 1987). It is clear, however, that the concept could also be used in risk assessment to determine w3 (serious injury). Also, for serious birth defects, QALY could be combined with loss of life expectancy to give w5. The strong aversion to serious birth defects, which shows itself in the large cost awards of compensation to victims of proved medical negligence or harm from a drug, suggests that society would assign a large value, possibly unity, to w_5.

Permanent forced evacuation is undoubtedly a serious and feared consequence of major hazards, but no method of determining w_4 is yet available. There is evidence that people will refuse to evacuate even when there is risk of death or incapacity, though it seems unlikely that they would do so in the face of certain death. The value of w4 would therefore be chosen to be somewhat less than one. On the other hand, a study by Kelly et al (1983) shows that about 1000 man-years of evacuation are required to avoid one fatal cancer under a range of evacuation and return criteria. This suggests a rather low value of w4 is implicit in evacuation plans.

Accidents and disasters lead to other unpleasant 'minor' consequences to individuals such as temporary evacuation (this is very common), disruption (for example, the need to take iodate tablets), and psychological trauma. Although it might be possible to quantify some of these consequences, for many it would be very difficult. They could be totally ignored as contributing to a measure of societal risk, but would at least partly be included in the financial measures to be considered next.

3.2.2 Non-personal risks

A societal risk is meant to represent the risk to society as a whole of the realisation of a specified hazard. So far we have considered it to be measured using a sum of the personal risks of the individuals affected. An obvious example, Three Mile Island (TMI), suffices to show that this measure is inadequate. In this accident the number of people affected is practically zero, yet no-one would deny that TMI was a major accident. There was considerable psychological trauma from temporary evacuation and other factors, but this is difficult to quantify. In financial terms, however, it assumes its true proportion: the cost of dismantling and cleaning up the plant alone has been put at about $1 billion. This, and the costs of the temporary evacuation and the replacement of electricity are costs which have ultimately to be met by society.

The major cost factors for a nuclear accident are:

- evacuation,
- resettlement,
- cleanup operations,
- interdiction of food,
- provision of uncontaminated water,
- sterilisation of land and property, and
- replacement of lost electrical generation capacity, possibly by more expensive means.

All these can, in principle, be quantified; estimates of the direct financial loss to the USSR due to Chernobyl range from $3-5 billion (Flavin, 1987). These costs are real, whereas the cancers thought to be induced probably cannot be observed against the normal background of cancer incidence.

The nuclear industry is not unique in being exposed to such cost risks; the Seveso incident involving dioxin is an example of one which could affect the chemical industry.

The total cost, C, of overt societal measures, as above, could be used to specify another measure of societal risk. For large nuclear accidents, or other major accidents which lead to extensive and prolonged contamination of the environment, it is at least as important to consider this financial measure of risk as risk measured by the number of real or hypothetical deaths. The financial measure incorporates evacuation costs which give some measure of the dominant disruption to individuals. The estimates for Chernobyl exceed $1 billion (Flavin, 1987).

To compare to the other personal risks, a total cost could be defined as

$$T = C + bN$$

where b is a value assigned to a human life.

Such an assignment is controversial and will not be pursued here except to note that, for nuclear accidents, b might have to be very large for the total not to be dominated by C. The question of the cost per unit of collective radiation dose to be applied for the purpose of cost benefit analysis in various situations is discussed in Appendix 1 where the point is made that cost per unit of risk is a well established principle which, although implying a cost per life, is a somewhat different concept.

For minor accidents, the direct costs may be negligible, but the indirect costs to an industry from loss of public confidence can be large. These might include the installation of expensive additional equipment, as at Sellafield, or a partial nuclear monitorium (Cave, Kastenberg and Tweedy, 1986). The frequency of minor accidents may be unimportant in contributing to personal and financial risks but may nevertheless be important to the industry concerned. This is an example of a non-personal risk which is not really financial in nature, but political. Because it is in the public eye, the nuclear industry is particularly liable to this type of risk. As we pointed out in Chapter 2, political aspects lie outside the scope of this report, but risk managers need to bear them in mind, nonetheless. Thus it is desirable to measure this type of risk in some way even though such considerations may not lead to actual gains to public safety.

3.3 Aspects of Nuclear Risk

In this section we discuss two matters which relate to the risk caused by nuclear installations. These are the effects of radiation and the question of using release as a surrogate for risk.

3.3.1 Risks from Radiation

It is not the intention of this subsection to give a detailed account of the health effects of radiation on individual organs, how these are assessed and what numbers are appropriate for use in risk estimation. We shall use National Radiological Protection Board (NRPB) advice on this, and the details are given by Pochin (1983), for example. But we do wish to mention some aspects of the quantitative representation of risk due to radiation, both individual and societal, which have the capacity to cause confusion and which are important for the risk comparisons discussed in Chapter 6. These arise largely from the stochastic (or random) nature of some effects of radiation.

Early effects are essentially non-stochastic. A dose of 4 Gy to the bone marrow from external radiation gives roughly a 50% probability of death and the range over which this probability goes from 0 to 1 is relatively small. This residual stochastic element is unimportant and it follows that the individual risk of early death is the frequency of accidents leading to doses over the threshold and the societal risk combines this over individuals. The situation is comparable to that for other types of fatal accident. Serious incapacity (for example due to lung fibrosis) is also effectively non-stochastic and depends mainly on the radiation dose.

In contrast the induction of cancers is stochastic; that is, the severity of the effect is independent of dose which determines only the probability of developing the effect. Genetic defects are also stochastic and can be treated similarly to cancers. For delayed death, the following complications are present (USNRC, 1985):

- many different types of cancer can be induced,
- there is a latency period during which the probability of cancer induction is effectively zero,
- following this latency period the probability of cancer induction may depend on the existing risk, or have an absolute value,
- the probabilities are assumed to be linearly proportional to dose at low doses (the so-called linear hypothesis); some models add quadratic terms,
- all these parameters may be a function of age, age at exposure and particular groupings such as gender, occupation, and so on,
- they are all rather uncertain but depend on the dose rate and
- the chance of death being caused by radiation depends on the probability of death in the intervening period due to some other cause.

To incorporate all these factors it is necessary to consider a quite complex and general model. There are two basic outcomes of the model for the individual. The first is the individual risk of death following an exposure. This is the probability that eventual death is caused by the exposure, or, to put it another way, that premature death is caused. This is a function of age at exposure, dose received and other groupings such as gender and occupation, though in practice models differentiate only on the basis of gender. It is, of course, also a function of the other risks to which the individual is exposed, and this is an important point to bear in mind when making risk comparisons. Thus a specification of maximum individual risk has to take these factors into account, and the effects can be large as will be seen in Chapter 6.

The probability that eventual death is caused by radiation takes no account of the fact that death is delayed. Thus the loss of life expectancy is considerably less than for immediate death, and this is another important representation of the risk from radiation. This, too, is a function of the various factors important for individual risk.

Using models of the hazard from radiation exposure (for example the recent Harvard models (USNRC, 1985)), it can be shown that the dependence of both these measures on dose is approximately linear, with the constant of proportionality for each individual depending on the factors we have identified above. However, it should be remembered that the effects have not been observed at low doses, and below some dose level the effects are simply unobservable. We return to the problems this creates for risk management in Chapter 9.

Because of the variations between individuals, combining these effects to get a societal measure depends on the age, group and existing risk characteristics of the particular society. In particular, the expected number of deaths, N is proportional to the collective dose under the linear model. It is the resulting constant of proportionality which is normally used in considering latent cancer risk: its ICRP (1977) value is 0.0125 per sievert. This is increased to 0.0165 if genetic effects are accounted for. Larger values are implied by the recent recommendations of the NRPB (1987). The previous discussion shows that by itself this number does not necessarily give a good description of the social impact of low levels of radiation since for some individuals the risk may be much higher than that suggested by the single constant.

The previous discussion refers to the expected number of deaths, $\bar{N}$; the actual number of deaths, N, is an (unobservable) random variable, which takes integral values. Consequence calculations (see Chapter 5) normally plot $\bar{N}$ not N and this can cause confusion since $\bar{N}$ can take any value, including fractions less than one. This also has a bearing on the next section where a distinction between discrete and continuous consequences is drawn. The precise form of the distribution of N is discussed by Kelly and Hemmings (1984).

Another way to represent the social risk would be to sum the loss of life expectancy over all individuals. As previously discussed, this could be used to give a weighting factor, w3, to compare the impact of early and delayed death.

3.3.2 Relation between releases and consequences

The definition of risk in Chapter 2 mentions 'specified undesired events'. So far in this chapter these events have been taken as human injury, damage to property or damage to the environment, but many probabilistic safety targets have used other types of event: system failures, plant damage, releases of radionuclides and doses to individuals. These quantities have the advantage of being easier to use and to demonstrate compliance with the targets, and are less uncertain. Of course they have the disadvantage that they do not directly address the events of concern. In this sub-section we examine a class of safety goals which has been used in the UK: these are based on release or dose.

The so-called Farmer line (Farmer, 1967) has been part of UKAEA risk management policy for many years. This is a line relating frequency and the release, R, of radioactive material into the environment, rather than actual harm to individuals or society; it is shown in Figure 3.2. A significant point about this line is that it relates to individual accident sequences, as we shall discuss in the next section. The important case, as far as effects on people are concerned, is when the release is into the air. When societal risks measured by costs, C, are concerned, it might also be necessary to include releases into other media - fresh water, sea or ground.

The NII's Safety Assessment Principles (HMNII, 1979) and the CEGB's Design Safety Criteria (CEGB, 1982a) also refer to the frequency of single accident sequences, and these frequencies relate to the maximum individual dose from the accident. This is a concept intermediate between release and individual risk. Note that when the the dose-risk factor is applied, it is not same as individual risk since for any individual allowance must be made for the fact that only a fraction of the time (about one in ten) is a given individual exposed to a dose close to the maximum. In practice, this concept is taken to have a direct relationship to release via a conservative set of assumptions about weather conditions.

Further details of all these approaches are given in Chapters 7 and 8, but it can be seen that the practice of defining safety targets in terms of releases, or closely related quantities, is well established in UK nuclear practice. We examine here how this relates to risk.

There are four reasons why release is not a good indicator of risk.

Single Measure of Release Farmer uses equivalent becquerels of I-131 to define R. This is not satisfactory over the broad range of release types which are possible in the nuclear industry. It is emphasised in Chapter 5 that releases have many aspects, including all the different radionuclides, temperature, release height and so on. This is why the NII and CEGB approaches refer to dose.

Site Dependence The harm to people and property will be dependent on the characteristics of each site. This is illustrated in the work of Gronow & Kelly (1984).

Weather Dependence We could expect that the collective dose is proportional to release, and hence, using the model in the previous subsection, that the expected number of delayed fatalities is also proportional to R. However, because of weather variability, this only applies to the average value of the collective dose, and there will be large fluctuations in practice. A particular result of this is that it is not obvious how to determine the relative frequency of high consequence accidents (as a result of 'high consequence aversion', say, see next section) by adjusting release frequencies, as Farmer attempts to do.

Non-Linear Effects There are other effects which mean there is a non-linear relationship between release and consequence; they relate both to the nature of the release, and the type of risk. For example we could compare the release of a cloud containing plutonium with one containing long-lived caesium. In the latter case it is absolutely necessary to take account of countermeasures which would inevitably be enforced; otherwise totally unrealistic risks would be calculated. As a result, although there may be approximate linearity between R and the average number of deaths (averaged over both the weather and the distribution of individual risks), for the plutonium release, this would certainly not be the case for the caesium release. Turning to risk type, it is clear that again countermeasures will cause non-linearities; they affect the number of health effects, as just described, and increase the financial costs at a super-linear rate as a result. Figure 3.3 shows some of these effects taken from the NRPB calculations for Sizewell 'B' (Kelly and Clarke, 1982). This figure shows a linear relationship for delayed cancer, with the remaining, economic, consequences increasing faster, some with thresholds. The same is true for early death.

It is therefore concluded that the relation between the release and risk is not simple, except possibly for delayed cancer, and even this will be site dependent. Thus safety targets should not be set in terms of release, though we expect it would be necessary to interpret risk targets in terms of releases for specific sites and radioactive inventories.

Because of these problems, some countries are considering targets put in terms of dose, taking account of weather conditions rather than simply assuming a prescribed, adverse weather with an individual on the plume centreline. Such targets avoid several of the shortcomings mentioned previously because they are more closely connected with the effects of true concern. However they cannot on their own address societal risk. This point is taken up again in Chapter 9.

3.4 Presentation of Risks

Because societal risk relates to different sizes of accident, and to the numbers of people suffering given levels of harm, there are a number of ways in which it can be represented in tabular, graphical or functional form. In contrast, individual risks are simply numbers and tables are sufficient to give all the information on total risks from different hazards.

Subsection 3.4.1 considers the basic representations in terms of frequencies, and subsection 3.4.2 assesses their use for safety goals. Subsection 3.4.3 looks at the distribution of risk among individuals in a population.

This section goes into some detail. The reason for this is that the different representations have been confused in the past, with like not being compared with like. A particular problem is the nature of safety goals when set in terms of the different representations. This can lead to plants complying with targets set in one form, but failing to comply with the equivalent target in another form.

3.4.1 Frequency distributions

The basic expression of a societal risk is a graph which shows the frequency of a given level of harm plotted against the level of harm. The level of harm may be either a discrete variable (number of deaths) or a continuous variable (expected number of delayed deaths, release, area of land contaminated to a given level). Since the discrete case is likely to extend up to large numbers this case can generally be treated as if the consequence measure were continuous. For the same reason the plots considered are generally logarithmic both in frequency and consequence. A number of ways of graphing societal risk are described in this subsection, and their use for safety goals is discussed separately in subsection 3.4.2.

Five representations which might be useful are identified; the first four are as follows.

Accident sequence scatterplots A set of points {fi,Ci} in the frequency-consequence plane, each one corresponding to a single accident sequence which must be defined in some way. An example is shown in Figure 3.4. The problem of a suitable definition is difficult, especially if we are concerned with a single consequence value. For example 'release' may be essentially a single number following a particular accident in a nuclear reactor, whereas 'collective dose' is not since it will depend on weather conditions. Since release is not a good measure of risk, as previously discussed, this method of representation has severe limitations.

Histograms The consequence axis is split into intervals and the frequency of consequences within each interval is plotted. An example is shown in Figure 3.5: the frequency of earthquakes in Iran. This figure shows that the dominant risk is from large earthquakes which kill very large numbers of people. This is because the large consequences outweigh the somewhat lower frequency of large earthquakes.

Frequency densities If C is a continuous variable, reducing the size of the intervals results eventually in a continuous curve with the interpretation that fd(C)dC is the frequency of events with consequences in the range C to C+dC. The corresponding representation for discrete consequences is simply the frequency at each consequence level. These representations are not recommended because they can be confused with each other and furthermore the density could either be per unit consequence, or per unit log consequence.

Complementary Cumulative Distribution Functions (CCDFs) These plot F(C) against C where F(C) is the frequency of events with consequences greater than or equal to C. An example is shown in Figure 3.6 which shows for a PWR that the frequency of fatal accidents, that is, one in which one or more people are killed, is $5x10^{-7}$ per year, and that the frequency of an accident killing 100 or more people is $1.5x10^{-7}$ per year. It is thus related to the density $f_d(C)$ defined above by:

$$F(C) = \int_C^{\infty} f_d(c)dc.$$

This representation has a number of advantages:
- it is invariant to transformations (unlike densities) so that logarithmic plots do not cause problems,
- the expected consequence per unit time is simply the integral of F:

$$\bar{C} = \int_0^{\infty} F(C)dC = \int_0^{\infty} Cf_d(C)dC$$

- such plots are also available for situations where the consequences are discrete. This is especially useful for plotting societal risk data as in Chapter 4.

CCDFs are commonly known as F-N (f-N, fN, f/N,...) lines (since N is used instead of C where the specific consequence is the number of people affected). Some further examples are shown in Figures 3.7-8 and these illustrate the range of types of risk which can be represented in this way. CCDFs have become the most usual way of representing societal risk, and we shall mainly be using them.

The expected consequence per unit time is a single number measure of societal risk used, for example, by Layfield (1987) in the report of the Public Inquiry on Sizewell 'B'. It corresponds to the "frequency x consequence" view of the risk due to a single accident. In terms of the scatterplot representation of a set of accidents with frequency f_i and consequence C_i, the CCDF is

$$F(C) = \sum f_i \text{ (summed over all i with } C_i \geq C)$$

from which it can be shown that the expected consequence, given by $\bar{C}$ above is

$$\bar{C} = \sum f_i C_i \text{ (summed over all i)}$$

which shows that the sum of frequency times consequence over all accidents is the above single risk measure. This single number removes a considerable amount of relevant information, for example, the frequencies, however small, of accidents with large C ($\gg \bar{C}$). It may not, therefore, represent well the public view of the risk.

A fifth representation which conveys both the integral risk measure as well as the information about accidents of different sizes which make it up has been proposed by Wheatley (1982).

First Moment Cumulative Distribution (FMCD) The idea is introduced as the expected consequences per unit time due to accidents with consequences greater than or equal to C:

$$G(C) = \int_C^{\infty} cf_d(c)dc$$

$$= \sum f_i C_i \text{ (summed over all i with Ci} \geq \text{C}$$

It follows that the total integral risk, *C* is *G(0)* (=*G(1)* for discrete consequences) and can thus be easily derived from a plot of G. In terms of the CCDF this becomes

$$G(C) = CF(C) + \int_C^\infty F(c)dc$$

This measure has certain advantages when used for safety goals (see 3.4.2), but because of its complexity we do not recommend it for use other than in the most technical contexts.

In any case it is possible to determine the source of overall risk from CCDFs. The integral measure of risk is dominated on a CCDF by the region with maximum CF(C), that is the part of a logarithmic CCDF with slope -1. For example, the CCDFs taken from the US Reactor Safety Study (USNRC, 1975), Figures 3.6-8, show the following consequence levels dominating overall risk of various types: early death, 200; latent cancers, 30; relocation area 30 square miles; decontamination area, 300 square miles. The results for early and delayed death, which are rather surprising at first sight, are due to early death risk being dominated by the most severe accidents with unfavourable weather conditions, whereas the expected cancers are mainly due to much more frequent, but less severe, accidents. This gives an indication of the information available from CCDFs.

3.4.2 Risk targets

Each of the above representations can be used to formulate quantitative targets in the appropriate terms. The nature of the target depends on what representation is used (see, for example, Munera and Yadigaroglu (1980)), but we do not pursue the technicalities of this here.

The importance of individual accident sequences is that they have been used in connection with the release type targets of the UKAEA, NII and CEGB described in subsection 3.3.2. The details in each case are described in Chapters 7 and 8. The comparison is between a line, such as a Farmer line, and a scatterplot representation of 'risk.' We note that such targets do not provide limits on the total number of accidents (and it is not difficult to split any sequence into two) and cannot, therefore, determine risk. This is in addition to the problems mentioned in subsection 3.3.2

The use of histograms is in principle fairly straightforward: "the frequency of accidents with consequences between C_1 and C_2 should not exceed f." They have been used in interpreting discrete accident sequence targets using assumptions such as "total frequency .... in each dose band is 10 times the assessment level frequency for each discrete fault sequence" (Harbison and Kelly, 1985). The results of this particular interpretation are shown in Figure 3.9 and they will be referred to at several points in the ensuing chapters. Interpretations of this kind require large numbers of technical assumptions and these are described by Harbison and Kelly, including the details of Cases A and B.

Frequency densities have not been used as safety goals, and, in view of the difficulties referred to in the previous subsection, this is not recommended.

CCDFs are very suitable for use as safety goals, since they minimise the difference between discrete and continuous consequences, they are reasonably easy to understand and they can represent 'high consequence aversion' in a simple way. As we noted previously, one measure of risk is the product of frequency and consequence, and so a target can be made 'high consequence averse' by having a slope steeper than -1 on a logarithmic plot. For example the provisional Dutch criteria described in Chapter 7 and shown in Figure 3.10 have a slope of -2. In this way the integrated risk implied by the target will be lower from high consequences than for low consequences.

We have also re-plotted the Iran earthquake data from Figure 3.5 on Figure 3.10. This comparison illustrates the large gap between the risk arising from a natural hazard in an undeveloped country and the aspirations of a safety conscious Western European country, though it should be remembered that the Dutch criteria apply only to a single plant.

All the above suggestions allow regulation to take account of the distribution of accidents leading to various sizes of consequences. If this is not desired, then the overall, integrated risk measure could be used. This can be made 'high consequence averse' by suitable weightings. For example we could replace a requirement on C by one on

$$S = \int_0^\infty C^{1+a} f_d(C)dC = (1+a)\int_0^\infty C_a F(C)dC \quad (a \geq 0)$$

This is the risk measure, C, previously described if $a=0$, and is 'high consequence averse' for $a>0$. This implies a tradeoff between the various consequence levels, with higher values assigned to higher consequences. This is completely different to the situation with CCDFs where the target has to be met at each and every consequence level; non-compliance at one level, however small, can not, in principle, be compensated by compliance, by however wide a margin, at another.

The final representation which we considered is the FMCD; no proposals using this format have been put forward. This is made high consequence averse by having a negative slope (though there are differences in detail between this and a slope steeper than -1 in a CCDF). The advantage for regulation of the FMCD when compared with the CCDF is that with CCDFs it is possible to change a non-compliant plant into a compliant one by replacing one set of accidents with another which has both higher overall risk, and higher consequences. In view of 'high consequence aversion', this appears to be an undesirable property which FMCDs do not have.

3.4.3 Relation to individual risks

The characterisation of societal risks by a frequency of N people being affected is rarely complete because in most cases the risk will not be spread uniformly over the population as a whole. The most obvious factor which removes uniformity is geographical: the proximity of an individual to the hazard. To describe this aspect of risk the concept of a risk contour has been introduced to relate the total risk to risk to individuals at a given distance from the source. An example of risk contours calculated for explosions at a coastal gas (SNG) plant in Scotland is shown in Figure 3.11.

One of the main personal risks following a nuclear accident is that of subsequent death from cancer, as discussed in subsection 3.3.1. A characteristic feature of such an accident is that large numbers of people acquire an additional very small risk of death from this cause. This feature has been graphically illustrated by

Grist (1982) who plotted the number of people, N(I), exposed to an individual risk I, or greater, against I, for a notional nuclear accident. Here I is the risk of death due to radiation induced cancer conditional on the occurrence of the accident; it is thus equivalent to dose. An adaption of his plot is shown in Figure 3.12, where the natural incidence of cancer mortality, a probability of 0.21, is also indicated. At the top end the plot shows that about 600 individuals acquire an additional probability of less than 1% of dying of cancer. At the lowest values of I, the plot can be cut off at any desired point corresponding to a fraction of the risk from natural background radiation, or a breakdown in the linear hypothesis. The use of this type of plot is implied by a recommendation of the Radiological Protection Commission in Germany (SSK, 1985).

The graph shown in Figure 3.12 is a CCDF, but here the consequence is individual risk, and it is the numbers affected that are plotted, not frequency. The expected number of additional deaths from the population exposed to risk in the range (I,I+dI) is IN(I)dI so that the total expected number of additional deaths among those exposed to a risk greater than I_L is

$$N = N(I_L)I_L + \int_{I_L}^{I_U} N(I)dI$$

where I_U is the upper limit to individual risk corresponding to a dose which is immediately fatal. From the shape of the curve, and this expression, it can be deduced that the dominant contribution to this comes from individuals whose risk is in the range $2x10^{-5}$-10^{-3}.

Finally we mention a related representation of the risk to a particular individual, which is a cumulative breakdown of I by accident. Thus H(I) would be the frequency of accidents for which the individual risk (or equivalently dose) conditional on the occurrence of the accident is equal to or greater than I. It is somewhat anomalous that traditionally social risk is represented in terms of accidents of differing severity whereas individual risk is aggregated over all accidents.

3.5 Conclusion and Recommendations

(1) Societal risk management should take into account at least five types of personal risk - early death, late death from cancer, serious injury, permanent or long-term evacuation and serious birth defects - as well as non-personal harm to society, which is principally financial.
(2) It is not appropriate for primary quantitative risk targets to be set in terms of a single simple release parameter.
(3) Of the many ways of representing risk, the CCDF (or F-N line) is the most accepted and the easiest to use and understand. Its use is therefore recommended for presenting the results of risk calculations. This does not necessarily mean that targets should be set using CCDFs.
(4) Although the integrated, single number, measure of social risk sometimes conceals important information, it can give a useful coarse description of the risk.

3.6 References

Cave, L, Kastenberg, W E and Tweedy, J N, 1986, **Some Possible Additions to the Value Term in Value Impact Analysis.**

CEGB, 1982a, **Design Safety Criteria for CEGB Nuclear Power Stations.** HS/R167/81 (Revised).

CEGB, 1982b, **PWR Design Safety Guidelines.** CEGB Generation, Development and Construction Division (DSG-2, Issue A).

Farmer, F R, 1967, **Siting Criteria - A New Approach.** IAEA Symposium on the Containment and Siting of Nuclear Power Reactors, Vienna 3-7 April, 1967. IAEA SM-89/34.

Flavin, C, 1987, **After Chernobyl: Reassessing the Costs of Nuclear Power.** European Environmental Review, 1(3), p 38.

Grist, D R, 1982, **Distribution of Individual Risk in Historical Data and in Plant Risk Assessment - A Personal View.** In 'Comparison of Risks Resulting from Major Human Activities', Xth regional congress of the International Radiological Protection Association, Avignon, France, 18-22 October, pp 51-58.

Gronow, W S and Kelly, G N, 1984, **Radiological Aspects of Site Selection for Nuclear Power Plants.** IAEA Safety Codes and Guides in the Light of Current Safety Issues. IAEA, Vienna.

Harbison, S A, and Kelly, G N, 1985, **An Interpretation of the Nuclear Inspectorate's Safety Assessment Principles for Accidental Releases,** IAEA Seminar on 'The Implications of PRA,' Blackpool, UK, March 1985. IAEA-SR-111/20.

ICRP, 1977, **ICRP Publication 27.** Annals of the ICRP, 1)4), p 1.

Kelly, G N, and Clarke, R H, 1982, **An Assessment of the Radiological Consequences of Releases from Degraded Core Accidents for the Sizewell PWR.** NRPB R137.

Kelly, G N, Ferguson, L and Charles, D, 1983, **The Influence of Countermeasures on the Predicted Consequences of Degraded Core accidents for the Sizewell PWR.** NRPB R163.

Kelly, G N, and Hemming, C R, 1984, **Probability Distributions of Radiological Consequences Following the Exposure of a Population.** Ann Nucl Energy, 11 (No 12), pp 597-606.

Kinchin, G H, 1978, **Assessment of Hazards in Engineering Work.** Proc Instn Civ Engrs, Part 1, 64, pp431-438.

Kind, P, Rosser, R, and Williams, A, 1982, **Valuation of the Quality of Life: Some Psychometric Evidence.** In 'The Value of Life and Safety,' M W Jones-Lee, ed, North Holland, Amsterdam.

Layfield, Sir Frank, 1987, **Sizewell B Public Inquiry Report.** HMSO, London.

Levine, S, et al, 1980, **Summary and Analysis of Safety Goal Proposals.** NUS Report NUS 3871 (Rev 1).

Marshall, W, Billington, D E, Cameron, R F, and Curl, S J, 1983, **Big Nuclear Accidents.** UKAEA Harwell Report, AERE-R10532.

Munera, H A, and Yadigaroglu, G, 1980, **A New Methodology to Quantify Risk Perception.** Nuclear Science and Engineering, 75, pp 211-224.

NRPB, 1987, **Interim Guidance on the Implications of Recent Revisions of Risk Estimates and the ICRP 1987 Como Statement.** NRPB-GS9, Chilton, Oxfordshire.

Pochin, Sir Edward E, 1983, **The Biological Basis of the Assumptions made by NRPB in the Calculation of Health Effects.** Sizewell 'B' Inquiry, Proof of Evidence, NRPB/P/2 (Rev).

Ramsay, C G, Sylvester-Evans, R, and English, M A, 1982, **Siting and Layout of Major Hazardous Installations.** I Chem E Symposium Series, No 71, pp 335-351.

Seaman, J, 1984, **Epidemiology of Natural Disasters.** In 'Contributions to Epidemiology and Biostatistics,' M A Klingberg, ed, S Karger, Basel.

Smith, A, 1987, **Qualms About QALYs.** The Lancet, May 16, pp 1134-6.

SSK, 1985, **Possibilities and Limits of the Application of Collective Dose.** GRS Safety Code and Guide 8/85.

USNRC, 1975, **Reactor Safety Study: An Assessment of Risks in US Nuclear Power Plants.** Wash-1400, NUREG-75/14.

USNRC, 1981, **A Study of the Implications of Applying Quantitative Risk Criteria in the Licensing of Nuclear Power Plants in the US.** NUREG/CR-2040.

USNRC, 1985, **Health Effects Model for Nuclear Power Plant Accident Consequence Analysis.** NUREG/CR-4214.

Versteeg, M F, and Visser, B J, 1987, **A PRA Guide for the Netherlands; a Consequence of the Dutch Policy on the Risk Management Applied to Nuclear Energy.** Presented at PSA'87, Probabilistic Safety Assessment and Risk Management Conference, Zurich.

Williams, A, 1985, **Economics of Coronary Artery Bypass Grafting.** Br Med J, 291, 326-329.

Wheatley, C J, 1982, **Conservatism and Bias in Societal Risk Criteria.** In 'Comparison of Risks Resulting from Major Human Activities,' Xth regional congress of the International Radiological Protection Association, Avignon, France, 18-22 October 1982, pp 51-58.

TABLE 3.1
Categories with Largest Numbers of People Affected

Event	Category with largest N_i	Category with second largest N_i
Bhopal	Serious incapacity, N_3	Early death, N_1
Chernobyl	Forced permanent evacuation, N_4	Delayed death, N_2

FIGURE 3.1
Disaster Profile of the Reactor Accident at Chernobyl

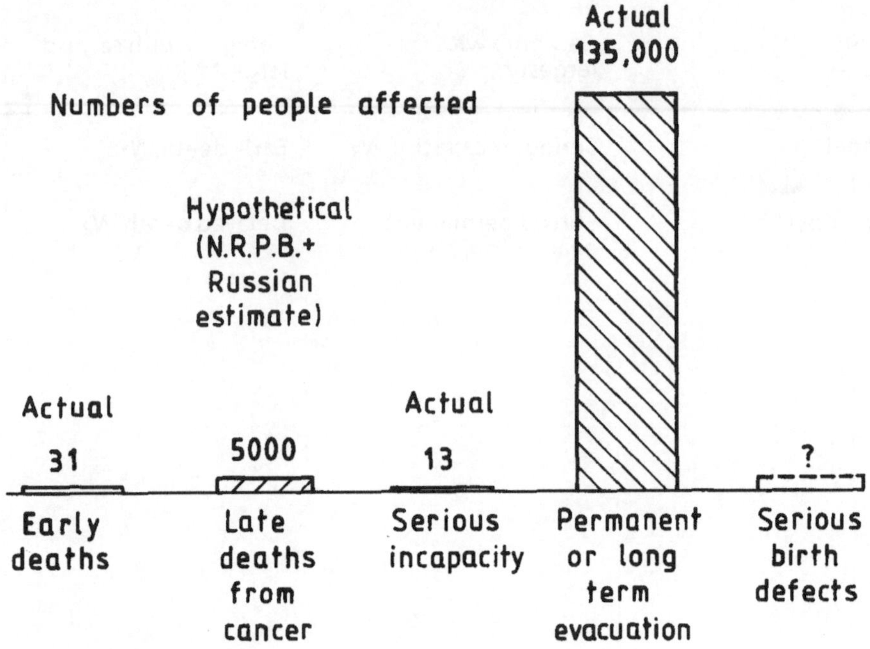

FIGURE 3.2
Farmer Release Criterion
(Source: Farmer, 1967)

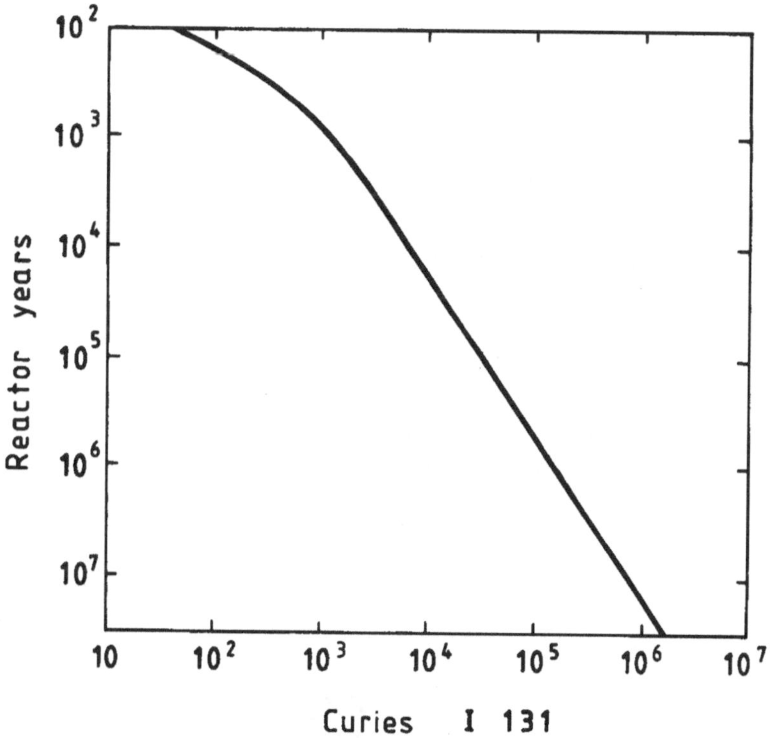

FIGURE 3.3
Sizewell 'B' Accident Consequences
(Adapted from Kelly and Clarke, 1982)

This shows the expected number of consequences (averaged over weather conditions) for various releases parameterised by *R*, the I-131 equivalent release defined by

$$R = \sum_i R_i(D_i / D_{I-131})$$

where D_i is the 50 year effective dose commitment (Sv) per Bq inhaled of radionuclide *i* and R_i is the release of *i* (Bq).

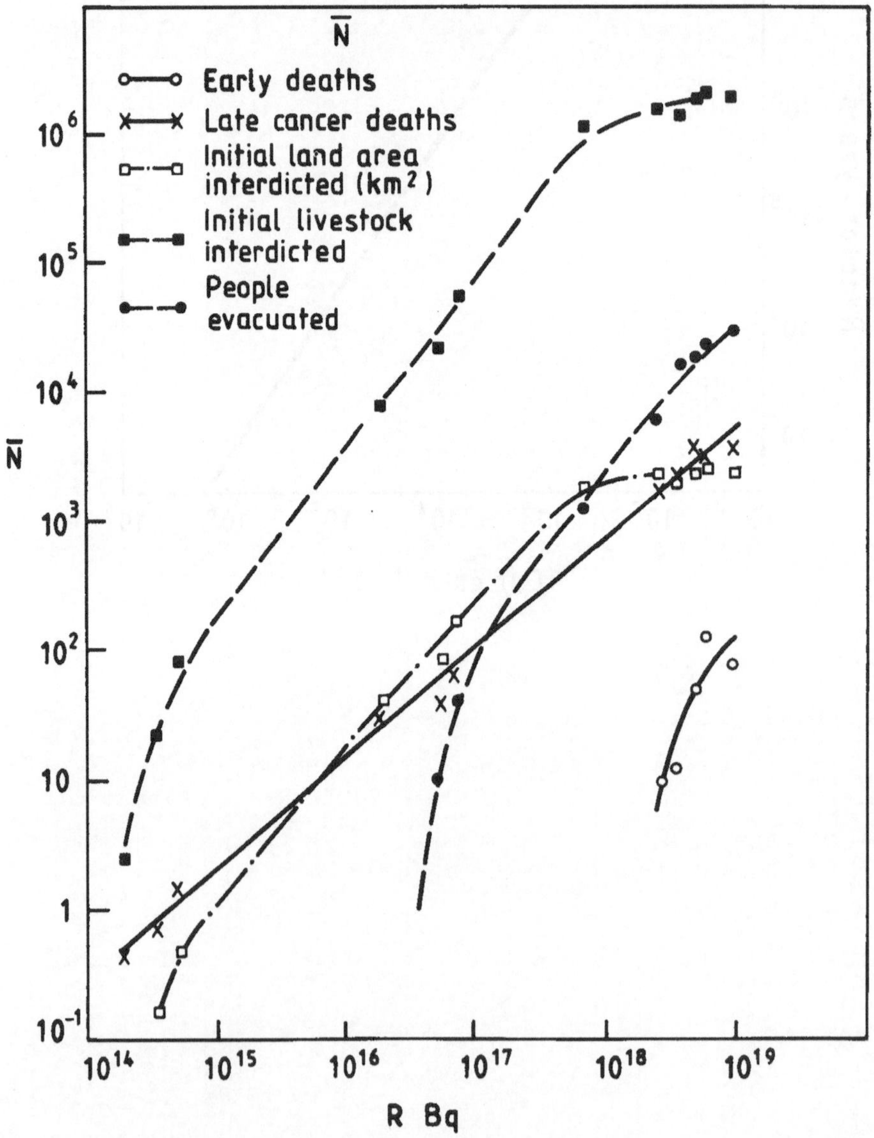

FIGURE 3.4
Example of Scatterplot: Accident Sequences in an AGR
(Source: Farmer, 1967)

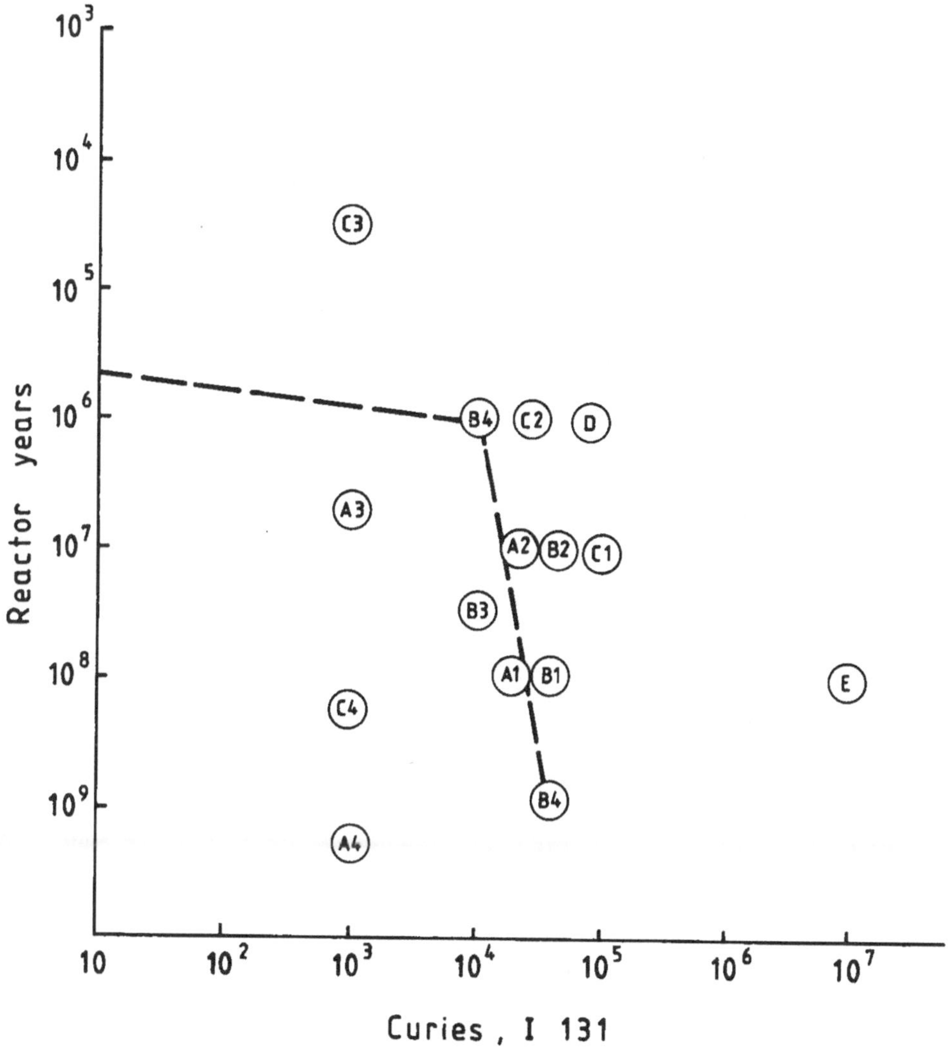

FIGURE 3.5
Example of Histogram: Earthquakes in Iran
(Adapted from Seaman, 1984)

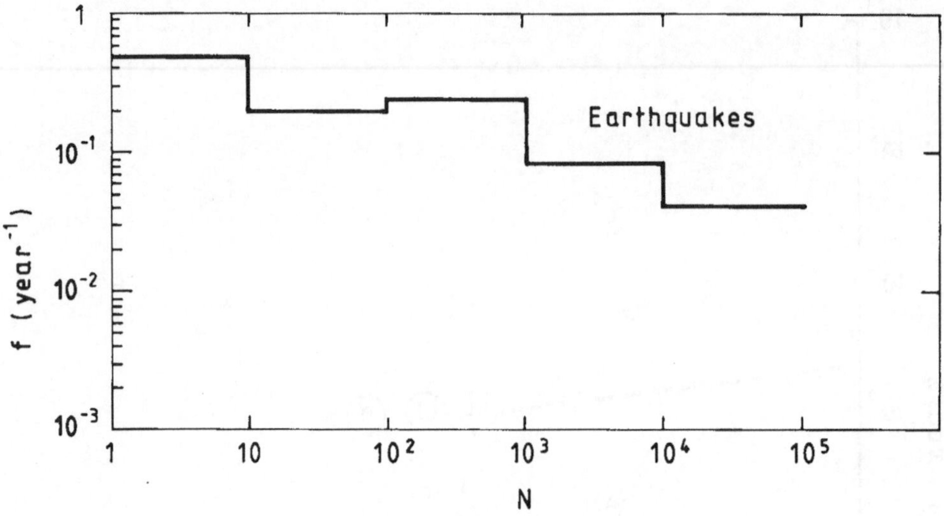

FIGURE 3.6
Example of Complementary Cumulative Distribution (1): Early Death in the Reactor Safety Study
(Source: USNRC, 1975)

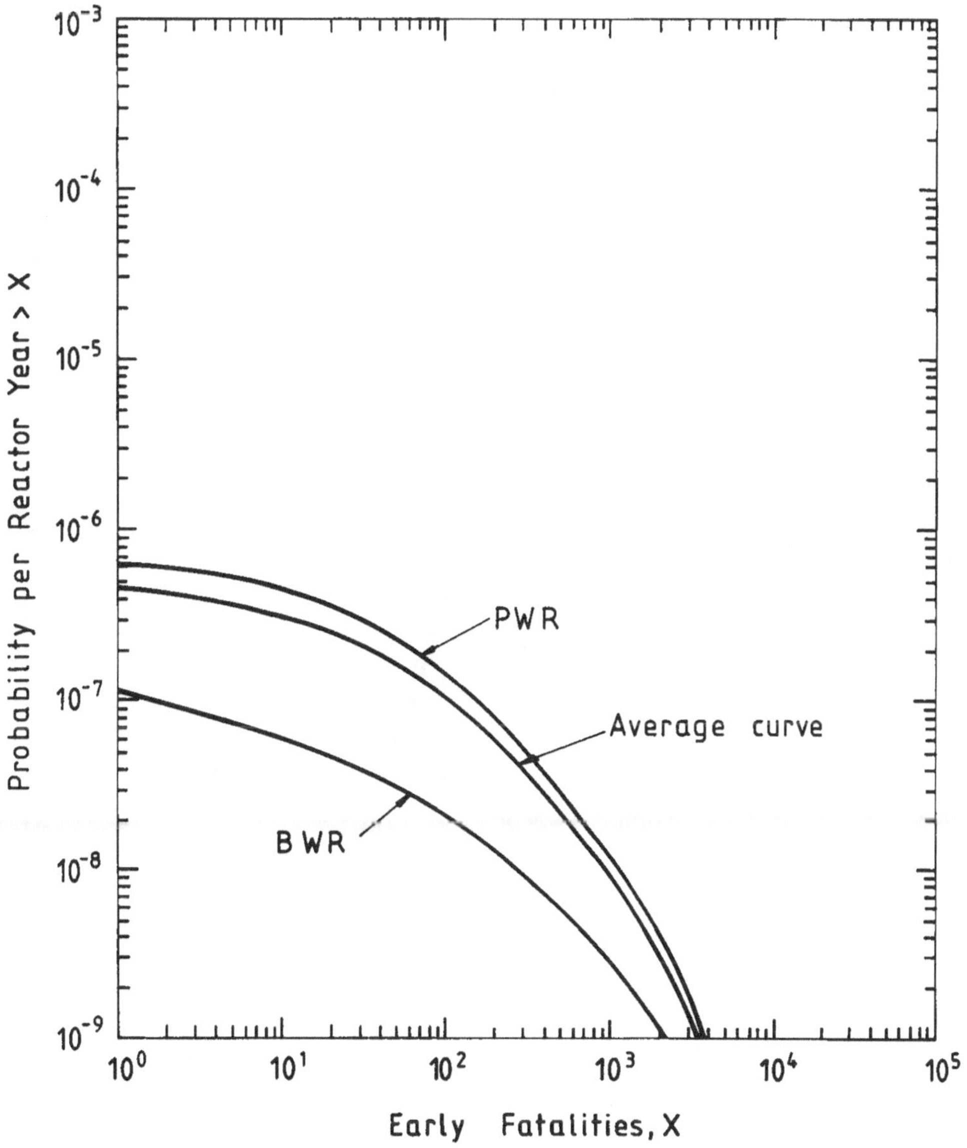

FIGURE 3.7
Example of Complementary Cumulative Distribution (2): Delayed Death in the Reactor Safety Study
(Source: USNRC, 1975)

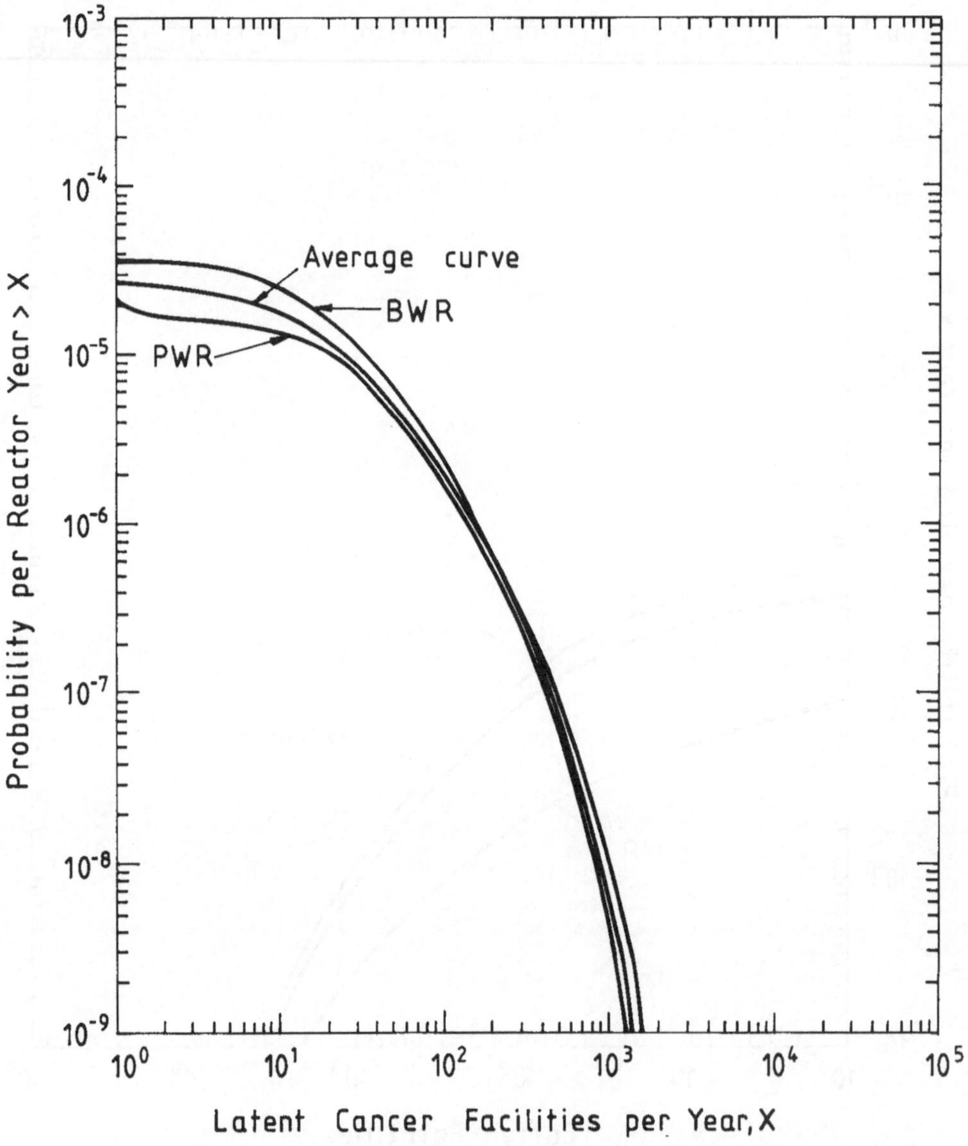

FIGURE 3.8
Example of Complementary Cumulative Distribution (3): Relocation and Decontamination in the Reactor Safety Study
(Source: USNRC, 1975)

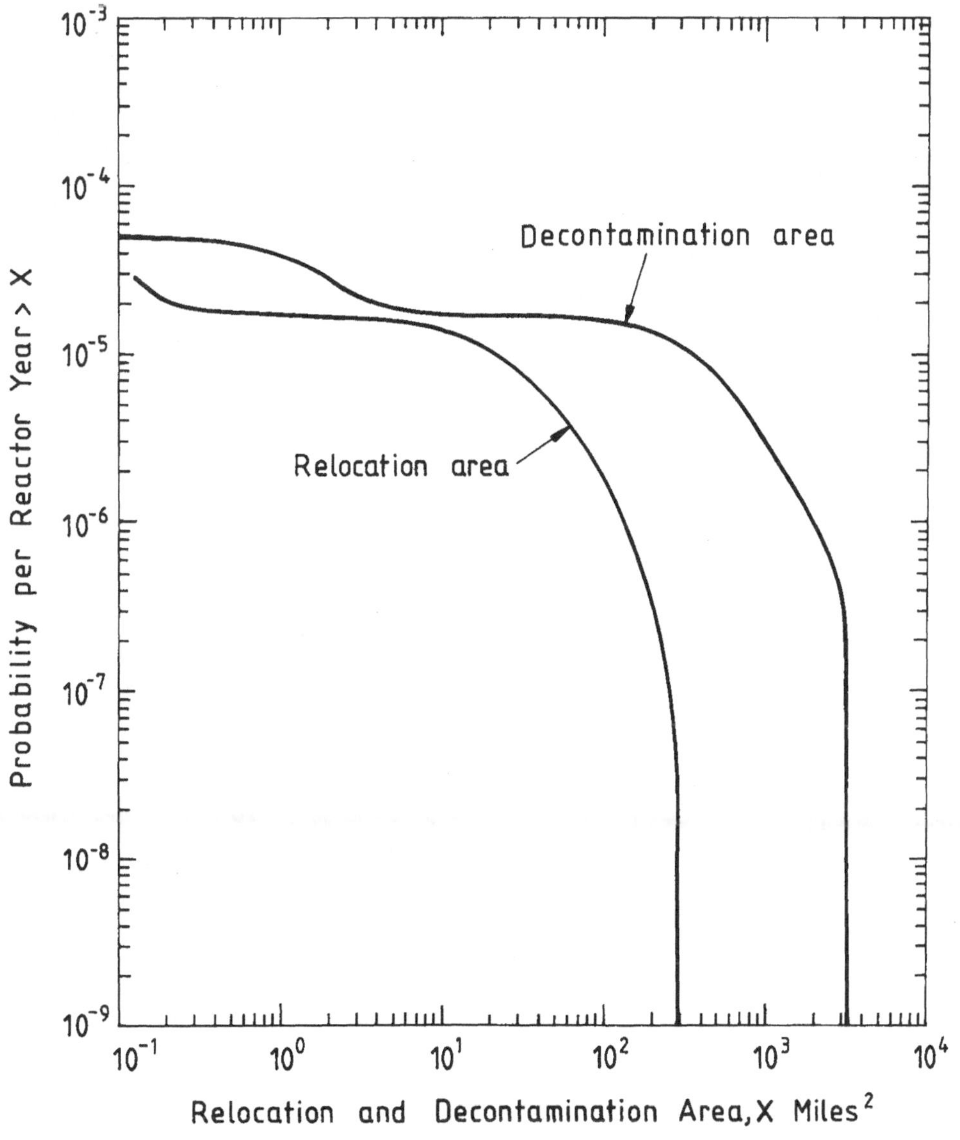

FIGURE 3.9
Histogram Interpretation of the NII's Safety Assessment Principles
(Source: Harbison and Kelly, 1985)

This shows accident frequency, *f*, against thyroid dose equivalent, *D*, for various dose bands in the NII Assessment Reference Levels. Cases A and B represent different assumed extensions of the principles above 1 ERL.

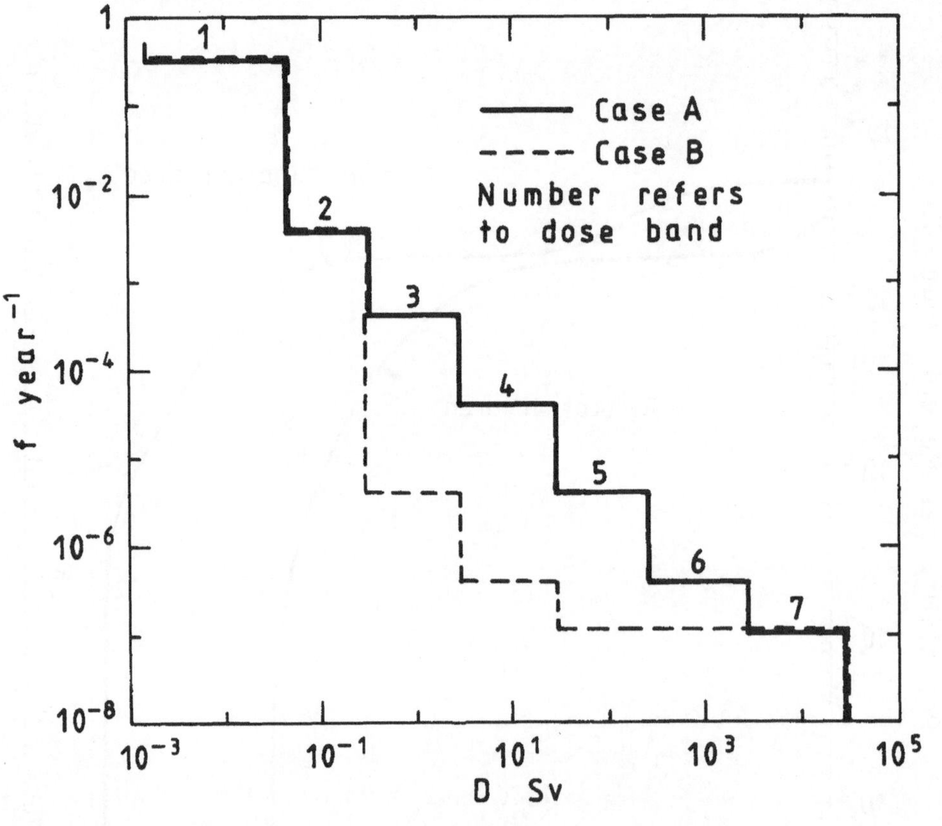

FIGURE 3.10
Provisional Dutch Safety Criteria: Early Death
(Adapted fromVersteeg and Visser, 1987)

This shows the Dutch CCDFs for early death (discussed further in Chapter 7) and also the Iran earthquake data re-plotted as a CCDF from Figure 3.4.

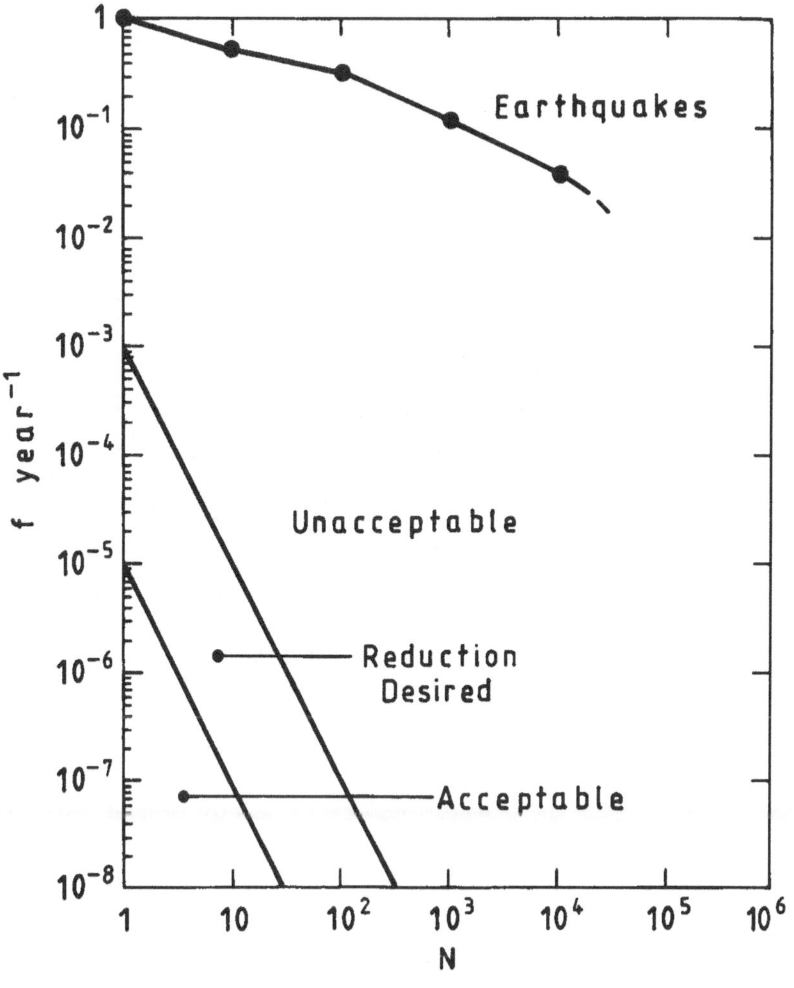

FIGURE 3.11
Example of Risk Contours
(Source: Ramsay et al, 1982)

Risk contours for an 0.03 bar explosion overpressure for the St Fergus gas reception terminal and SNG plant in North East Scotland. The figures are the frequency of this overpressure at each location in units of 10^{-6} per year.

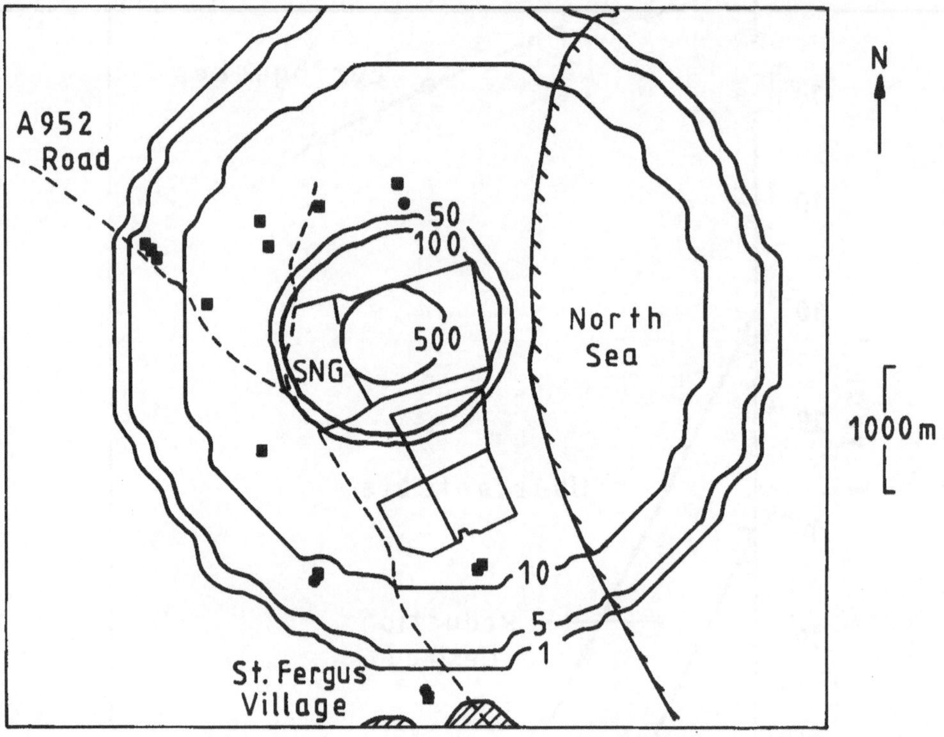

FIGURE 3.12
CCDF for Number of People at Risk
(Adapted from Grist, 1982)

Number of people, *N*, subject to an individual risk of *I* or greater, where *I* is the risk of death from cancer conditional on a high release accident at a notional nuclear site.

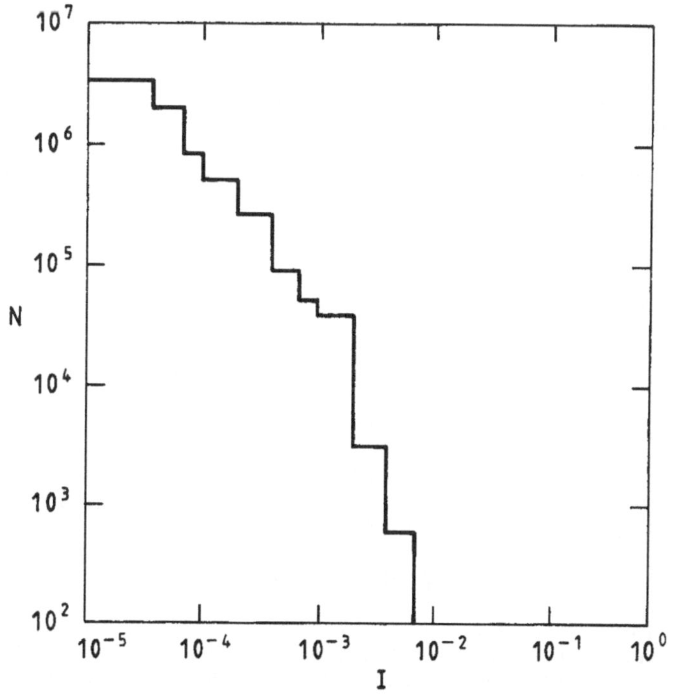

(probability of death from cancer conditional on the accident)

CHAPTER 4

EXISTING RISK

4.1 Introduction

No activity can be undertaken without some associated risk. In order to help to set risk results from probabilistic assessments of new technology in context, we use what historical data there are on existing risk arising from activities in society. This 'background risk' arises from accidents, the working environment, leisure activities and medical conditions.

Since the establishment of existing risk relies on historical data, our knowledge is reliant upon a slowly increasing awareness of the need to compile relevant statistics. Sadly, the occurrence of a catastrophic accident can often be the initiating event for a compilation of a full data base in a particular risk field. For this reason, historical data on existing risk are sparse, and have to be obtained from disparate sources. The literature presenting background risk statistics for the UK such as Grist (1975), and Fernandes-Russell (1987a) on individual risk; and Fryer & Griffiths (1979) and Fernandes-Russell (1987b) on societal risk, concentrate on certain types of risk, including death from medical conditions, industrial accidents and natural hazards. In contrast, very little data are available on the individual risk of injury in the general population.

There are several types of event for which the risk is not easily estimated using historical data. For some events the consequences are difficult to measure or detect (for example, cancer from background radiation). There are also cases for which the attribute measured as the consequence is not representative of the risk (for example, where an event produces more injuries than fatalities, but only fatalities are recorded). In addition, for events which have a low frequency of occurrence, the uncertainty in the risk estimate will increase as the frequency decreases. Taken to its extreme, this category includes events yet to occur. The risk due to high frequency events can be much more precisely determined than the risk due to relatively infrequent events. In Table 4.1 the high frequency of occurrence of road accidents can be contrasted to the much rarer occurrence rate of fatal accidents in the chemical and oil industries.

In the following two sections, illustrative values of the current risk of death for individual and societal risk are presented. Recorded data on risk of injury are relatively sparse; however, one table on occupational non-fatal injuries has been included. The values of individual risk presented here represent the fraction of a population who have died or sustained injury in one year. Risk measures such as the fatal accident rate (FAR), the loss of life expectancy and the number of deaths per unit of activity (such as passenger-miles) have not been included.

The individual risk measure chosen here is sensitive to the definition of the population at risk chosen for the calculation. This causes a difficulty in risk comparisons since individual risks for different categories are often evaluated for different populations and therefore an inequitable comparison can result. For instance the population at risk used in the calculation of occupational fatal injury risk (Table 4.6) is the total number of employees in each industry. This includes employees who are not directly exposed to industrial accidents, such as those who work in an administrative capacity. Care needs to be taken in comparisons between these figures and those estimated, for example, for radiation workers. The individual risk for the latter is usually estimated for the workforce which is directly exposed to risk (Hughes and Roberts, 1984). For the individual risk values presented here, the population is specified on each table.

Cumulative frequency versus consequence curves are presented from which a comparison of the current societal risk from various activities can be made. Again the frequency is estimated by dividing the number of events by the period for which data are available. This leads to higher uncertainty in the estimates at higher consequence.

4.2 Individual Risk

Table 4.2 displays the approximate individual risk of death for the British population as a whole from major causes. Although permitting broad comparison, these figures are of limited value since they represent the risk to an individual averaged over the whole population. For an individual in the most exposed group, the resulting risk estimate may be several orders of magnitude greater. This can be seen from Figure 4.1 where this risk is plotted as a function of age. Criteria other than age may be used to define the population at risk. These include geographical location, sex, occupation, or a readily identifiable attribute such as smoking in the case of lung cancer. Between each of these distinctions, the variation observed spans several orders of magnitude, demonstrating the sensitivity of the individual risk estimate to the defined population at risk.

As can be expected deaths due to internal causes (disease) rather than external causes (accidents and deliberate killings including suicide) dominate the mortality statistics. Some examples are displayed in Table 4.3.

In the context of nuclear risk management it is particularly important to examine death from cancer and accidental death. Figure 4.1 also shows how the risk of death from these causes varies with age and sex; these data are used in the comparisons in Chapter 6. The accidental death statistics are dominated by motor traffic accidents for young people (especially men) and by accidental falls for older people. This explains the two distinct peaks seen in the Figure. The risk of death from accidents due to a variety of causes is shown in Table 4.4.

It is important to note the systematic variation in the risk over time due to changing social and environmental factors. Figure 4.2 displays the annual number of fatalities in the coal mining industry from 1949 to 1987. The time dependent nature of the data can have important implications for those involved in specifying future risk targets based on values of existing risk.

The individual risk of death from fatal injuries occurring at work can be as high as 1.8×10^{-4} per year for the mining and quarrying industry. The construction industry also involves a high level of fatal risk. Values for a variety of industries are displayed in Table 4.5 which shows the risk averaged over all the employees in each industry and not just those who would be directly at risk from industrial accidents. For the latter group the risks would be higher.

In the UK, information on the rate of occurrence of injuries in the general population is not readily available. One of the main difficulties is the definition of injury severity. Data on occupational injury are, however, available for the years 1981-84 (Fernandes-Russell, 1987a). Table 4.6 displays the individual risk for major injuries for several industries. A "major" injury is defined as one which requires leave from work of 3 days or more. The figures do not solely include injuries directly associated with work but all injuries at the workplace.

The figures described above are of relevance in a comparison of early death risk. Due to the time lapse between exposure to the hazard and the development of the detriment, the risk of death from occupational disease is a better comparator to disease induced by radiation. However, just as for industrial injury risk, the risk of death due to occupational disease is difficult to quantify and monitor. This is due to problems in the definition of consequence severity and the long timescale over which a disease can develop. Furthermore, the risk of a particular occupational disease is usually greatly reduced once it is recognised. Some figures are quoted in Table 4.7 from Pochin (1975), but they cannot be regarded as representative of the present situation for the reasons just stated.

From the evidence to the Sizewell 'B' Public Inquiry, the average individual risk estimated for workers, that is, the combined total of occupational risk, fatal cancer risk and hereditary defects risk was 1.3×10^{-4} per year. The maximum individual risk to members of the public was estimated at 3.4×10^{-7} per year (Layfield, 1987). The risk associated with future generations is included in these values. This should be borne in mind when comparing these figures with the statistics of industrial risk which are normally computed on the basis of early death due to accidents, as well as the point that one is comparing the values of a formal estimate of future health risks against the historical record of actual accidents.

4.3 Societal Risk

As discussed in Chapter 3, the frequency-consequence curve or CCDF is the most common means of representation of societal risk. It has the feature of demonstrating the variation in frequency of occurrence of events according to the magnitude of the consequence. In CCDFs, the frequency, F, with which accidents leading to N or more casualties occur is plotted against the consequence, N. The advantages and disadvantages of the use of CCDFs for comparative purposes are discussed in Griffiths (1981). Here we are concerned with comparison of numerical values for existing societal risk through the use of this common measure.

The most widely used measure of societal risk is the frequency of occurrence of multiple fatality accidents. Figure 4.3 illustrates the frequency of multiple fatality accidents in the United Kingdom for selected man-made hazards. The points shown on this and subsequent figures relate to equal logarithmic intervals of consequence rather than actual events.

The frequency of occurrence is sensitive to the time period to which the data relate as a result of systematic changes over time. These include increases in technological safety, changes in consumer habits, increased efficiency of emergency services, changes in the technology employed (for example, larger planes), and changes in risk management. Figure 4.4 displays the change in the slope of the CCDF for worldwide aircraft accidents for two 20-year periods.

The number of fatalities, however, is not the only parameter that can be used to monitor existing societal risk from catastrophic events. As discussed in Chapter 3, the multi-attribute nature of catastrophic events requires special consideration. Aspects of societal risk other than multiple fatalities include the consideration of multiple injuries and social disruption such as the number of people evacuated, the area of land contaminated, loss of income and loss to the economy.

Figure 4.5 presents CCDFs for three types of consequence, death, injury and evacuation, for chemical and petrochemical industrial accidents in the UK and also worldwide. The number of people harmed refers to both off-site and on-site involvement. Such detailed information on the level of societal risk posed by a particular activity is not generally available for many activities. The chemical industry has been particularly well documented on safety issues. Another industry with a well documented accident record is the air transport industry. Figure 4.6 displays the CCDFs for death and injury for aircraft accidents that have occurred in the UK and worldwide.

A comparison of the societal risk posed by some natural hazards and man-made hazards for worldwide events can be made from Figures 4.7 and 4.8. Natural hazards can often involve casualties more than one order of magnitude greater than typical casualty levels for man-made hazards. The CCDFs for natural catastrophes appear to be 'risk seeking' rather than 'risk averse' in that most of the total number of deaths are due to high consequence events. In contrast the man-made hazards are approximately 'risk indifferent' (that is, slope -1 on a log-log plot) up to the maximum observed consequence. (This is, of course, a convenient abuse of these terms. It is safety targets which have an attitude to risk, not observed levels of risk.)

Arguments regarding acceptable or accepted levels of risk can use estimates of current risks posed to society in order to set a context from which to evaluate the results from probabilistic studies of new technology. These arguments will be developed in Chapter 6. In Figure 4.9, CCDFs from natural hazard events occurring in the US are set against those obtained from the PSA results of the Reactor Safety Study (see Figures 3.5-7 and Chapter 7). Figure 4.10 shows a similar graph in which U.S. man-made hazards are set against the same PSA curves. A more recent US Study (NUS, 1985) repeats these US calculations and examines the effect of high consequence aversion on the ranking of various hazards. These rankings are reproduced in Table 4.8. The capacity of dam failures to cause high consequence accidents is reflected in its jump from 10th to 3rd in order of severity when this is accounted for with a factor proportional to the size of the consequences. Figure 4.11 displays some UK information: the PSA results from the Canvey Island and Sizewell 'B' studies (see Chapters 7 and 8) involving casualties and death respectively. These theoretically derived figures present frequencies which are at least two orders of magnitude lower than the total event categories presented in Figure 4.3. The latter are common risks already accepted by society but they are not observed to extend up to the very high consequences shown by the predictions of formal risk analysis.

4.4 Summary of Key Points

(1) Existing levels of average individual risk in Great Britain are 10^{-2} per year for death from all causes, and is as low as $2x10^{-4}$ for the age group least at risk.

(2) The corresponding figures for death due to cancer in Great Britain are $3x10^{-3}$ per year and $4x10^{-5}$ per year respectively.

(3) The corresponding figures for death from accidents in Great Britain are $3x10^{-4}$ per year and $9x10^{-5}$ per year.

(4) The individual risk of death from fatal injuries at work in Great Britain can vary between $1x10^{-6}$ per year and $880x10^{-6}$ per year depending on the industry considered. The average over all industries is $22x10^{-6}$ per year.

(5) The individual risk of injury at work in Great Britain can vary between $0.1x10^{-6}$ per year and $28x10^{-4}$ per year depending on the industry considered. The average over all industries is $6x10^{-4}$ per year.

(6) The data used in the risk estimates are time dependent and the current trend is towards levels of increasing safety.

(7) The individual risk calculation is very sensitive to the choice of the population at risk.
(8) For many natural disasters most deaths take place in the highest consequence events observed whereas in man-made accidents they are evenly distributed through the consequence range, or are concentrated in low consequence events.

4.5 References

Barrell, A C, Edmondson, J N, and Holden, P L, 1985, **Canvey Island - A Case Study of the Application of PRA**. IAEA Seminar on the Implications of PRA, Blackpool, 18-22 March, 1985.

Bromley, J, 1987, **Comparing the hazards of coal and uranium mining**. Atom, No. 365, pp 3-9, March 1987.

Coppola, A, and Hall, R E, 1981, **A Risk Comparison**. NUREG/CR-1916.

Department of Transport, annual publication, **Road Accidents Great Britain**. HMSO, London.

Fernandes-Russell, D P, 1987a, **Individual Risk Statistics for Great Britain (1980-1984)**. Research Report No 2, Environmental Risk Assessment Unit, University of East Anglia, Norwich.

Fernandes-Russell, D P, 1987b, **Societal Risk Estimates from Historical Data for UK and Worldwide Events**. Research Report No 3, Environmental Risk Assessment Unit, University of East Anglia, Norwich.

Fryer, L S, and Griffiths, R F, 1979, **World-wide Data on the Incidence of Multiple-Fatality Accidents**. UKAEA Report SRD R149, HMSO, London.

Griffiths, R F, 1981, **Problems in the Use of Risk Criteria**. In Dealing With Risk, Manchester University Press.

Grist, D R, 1978, **Individual Risk - A Comparison of Recent British Data**. UKAEA Report SRD R125, HMSO, London.

HSC, 1987, **Executive Reports for 1986/7**. HMSO, London.

Hughes, J S, and Roberts, G C, 1984, **The Radiation Exposure of the UK Population - 1984 Review**. NRPB R173.

Institution of Chemical Engineers, annual publication, **List of Incidents, Loss Prevention Bulletin**. I Chem Eng, Rugby.

Kelly, G N, and Clarke, R H, 1982, **An Assessment of the Radiological Consequences of Releases from Degraded Core Accidents for the Sizewell PWR**. NRPB R137.

Kletz, T A, 1971, **Hazard Analysis, a quantitative approach to safety**. Symposium on Major Loss Prevention in the Process Industries, I Chem Eng, Newcastle-upon-Tyne.

Layfield, Sir Frank, 1987, **Sizewell 'B' Public Inquiry Report**. Chapter 47, pp 20 and 23.

National Coal Board, annual publication, **Annual Report.**

NUS 1985, **Development of Methodology for Comprehensive Hazard Analysis.** Study for US FEMA by NUS Corporation, NUS 4721.

Pochin, E E, 1975, **The Acceptance of Risk.** British Medical Bulletin 31(3), pp 184-190.

TABLE 4.1
Frequency of Road Accidents Compared to Accidents in the Chemical and Petrochemical Industries Great Britain 1981-1985

Year	Road Accidents No of accidents with *x* deaths *x*				Chemical and Petrochemical Industry Accidents No of accidents with *x* deaths *x*				
	1	2	3	4+	1	2	3	4	15
1981	4969	306	61	19	1	0	0	0	0
1982	5056	320	54	17	1	1	0	0	0
1983	4691	278	39	19	3	1	0	1	0
1984	4789	274	51	24	2	2	1	0	1
1985	4456	261	33	18	2	1	0	0	0

Sources: Department of Transport
Institution of Chemical Engineers

TABLE 4.2
Overview of the Individual Risk of Death for Broad Categories Great Britain 1980-1984

	Individual Risk (all ages) (per year)
All causes	10^{-2}
All internal causes (medical ailments and disease)	10^{-2}
All external causes (accidents, violence and poisonings)	10^{-4}
All accidents	10^{-4}
All industrial accidents	10^{-5}

Population at risk: total population in Great Britain.

Source: Fernandes-Russell (1987a)

TABLE 4.3
A Comparison of the Individual Risk of Death from Internal and External Causes Great Britain 1980-1984

	Individual Risk (all ages) (per year)
Internal Causes	
Diseases of the Circulatory System	5800×10^{-6}
Neoplasms	2700×10^{-6}
Diseases of the Respiratory System	1600×10^{-6}
Diseases of the Digestive System	300×10^{-6}
External Causes	
Motor Vehicle Traffic Accidents	100×10^{-6}
Accidental Falls	90×10^{-6}
Suicides	90×10^{-6}

Population at risk: total population in Great Britain.

Source: Fernandes-Russell (1987a)

TABLE 4.4
The Individual Risk of Accidental Death for Selected Causes
Great Britain 1980-1984

	Individual Risk (all ages) (per year)
Lightning	0.1×10^{-6}
Misadventure to patients during surgical and medical care	0.9×10^{-6}
Air and Space Transport	1.2×10^{-6}
Accidental Poisoning	13.1×10^{-6}
Motor Vehicle Accidents	101.0×10^{-6}

Population at risk: total population in Great Britain.

Source: Fernandes-Russell (1987a)

TABLE 4.5
Fatal Injuries at Work for Selected Industries
Great Britain 1981-1984

	Individual Risk (per year)
Professional and scientific services	$1x10^{-6}$
Electrical engineering	$6x10^{-6}$
Paper, printing and publishing	$11x10^{-6}$
Chemicals and allied industries	$19x10^{-6}$
Transport and communication	$33x10^{-6}$
Construction	$102x10^{-6}$
Coal underground mining	$150x10^{-6}$ (1)
Mining and quarrying	$179x10^{-6}$
Quarries	$390x10^{-6}$ (2)
Deep sea fishermen on UK vessels	$880x10^{-6}$ (2)
Metal manufacturing industry	$76x10^{-6}$
Average over all industries	$22x10^{-6}$

Population at risk: total number of employees in each industry except for (1) where it is the total number of miners.

Sources: Fernandes-Russell (1987a)
(1) Bromley (1987)
(2) HSC (1987)

TABLE 4.6
Non Fatal Major Injuries for Selected Industries
Great Britain 1981-1984

	Individual Risk (per year)
Professional and sicentific services	10×10^{-6}
Electrical engineering	310×10^{-6}
Shipbuilding	1070×10^{-6}
Timber, furniture, etc.	1500×10^{-6}
Chemicals and allied industries	940×10^{-6}
Construction	1950×10^{-6}
Mining and Quarrying	2800×10^{-6}
Average over all industries	600×10^{-6}

Population at risk: total number of employees in each industry.

Source: Fernandes-Russell (1987a)

TABLE 4.7
Occupational Disease

	Individual Risk (per year)
Beta-naphthylamine (bladder cancer)	24000×10^{-6}
Rubber mill workers (bladder cancer)	6500×10^{-6}
Underground mining (pneumoconiosis and silicosis)	$40\text{-}5800 \times 10^{-6}$ *
Viscose spinner (heart disease)	3000×10^{-6}
Coal carbonisers	2800×10^{-6}
Asbestos (lung cancer)	$2300\text{-}4100 \times 10^{-6}$
Uranium mining (lung cancer)	1500×10^{-6}
Wood machinists (nasal cancer)	700×10^{-6}
Shoe industry (nasal cancer)	130×10^{-6}

* Large differences in rates for different countries and different types of work.

Source: Pochin, 1975

The details, including the population involved, are unknown.

TABLE 4.8
Ranking of US hazards
(Source: NUS, 1985)

Unweighted Rank	Hazard	Weighted Rank
1	Extreme heat or cold	1
2	Aviation accident	10
3	Urban fires	7
4	Hurricanes	2
5	Tornadoes	4
6	Flood	7
7	Marine accident	4
8	Severe winter storm	7
9	Hazardous material incident	11
10	Dam failure	3
11	Railroad accident	13
12	Wild fires	6
13	Earthquake	12
14	Landslide	14
15	Tsunami	15
16	Avalanche	16

Note: the unweighted rank is based on mean consequences whereas the weighted rank incorporates a factor proportional to consequence to account for 'high consequence aversion'.

FIGURE 4.1
Individual Risk as a Function of Age and Sex
All Causes, Cancer and Accidents
Great Britain 1984

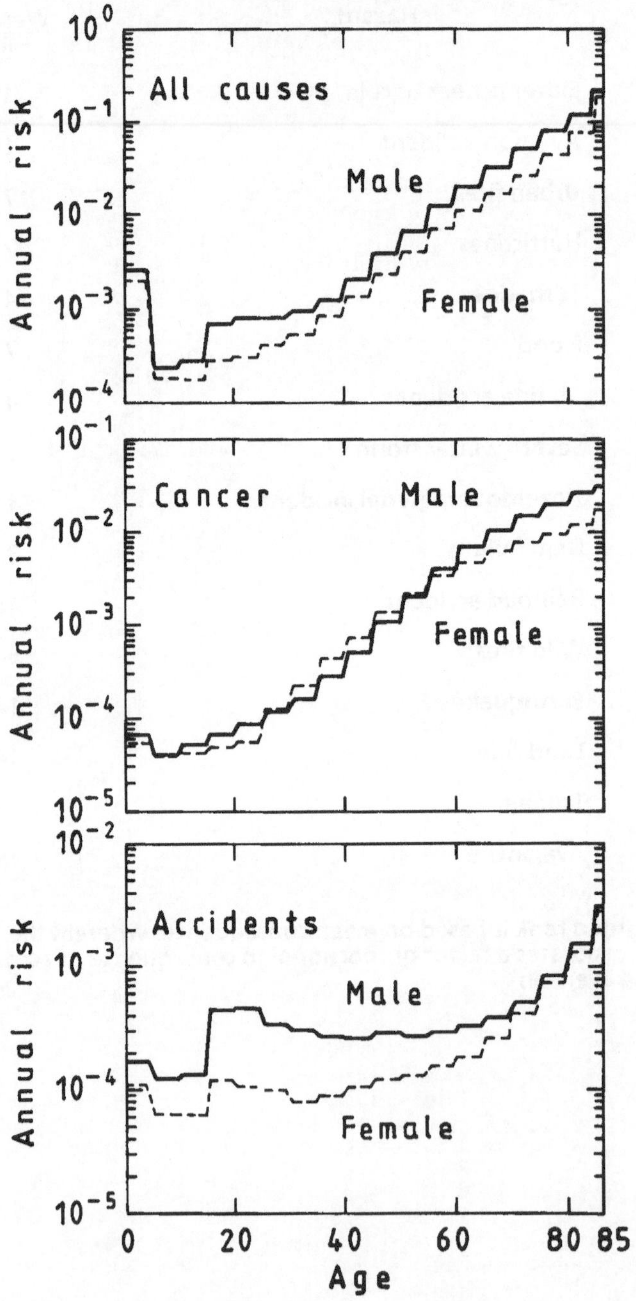

FIGURE 4.2
Fatal Accidents Occurring in Coal Mines
Great Britain 1949-1987
(Adapted from National Coal Board)

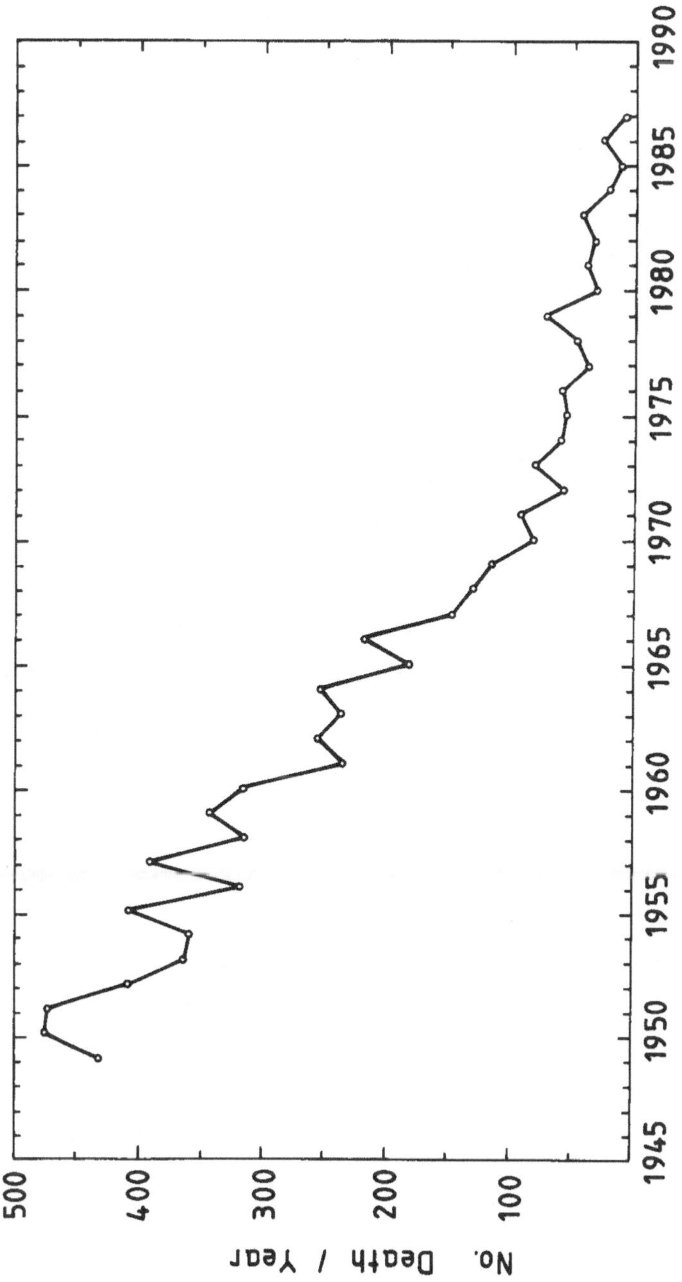

FIGURE 4.3
Frequency of Multiple Fatality Accidents in the UK
(Source: Fernandes-Russell, 1987b)

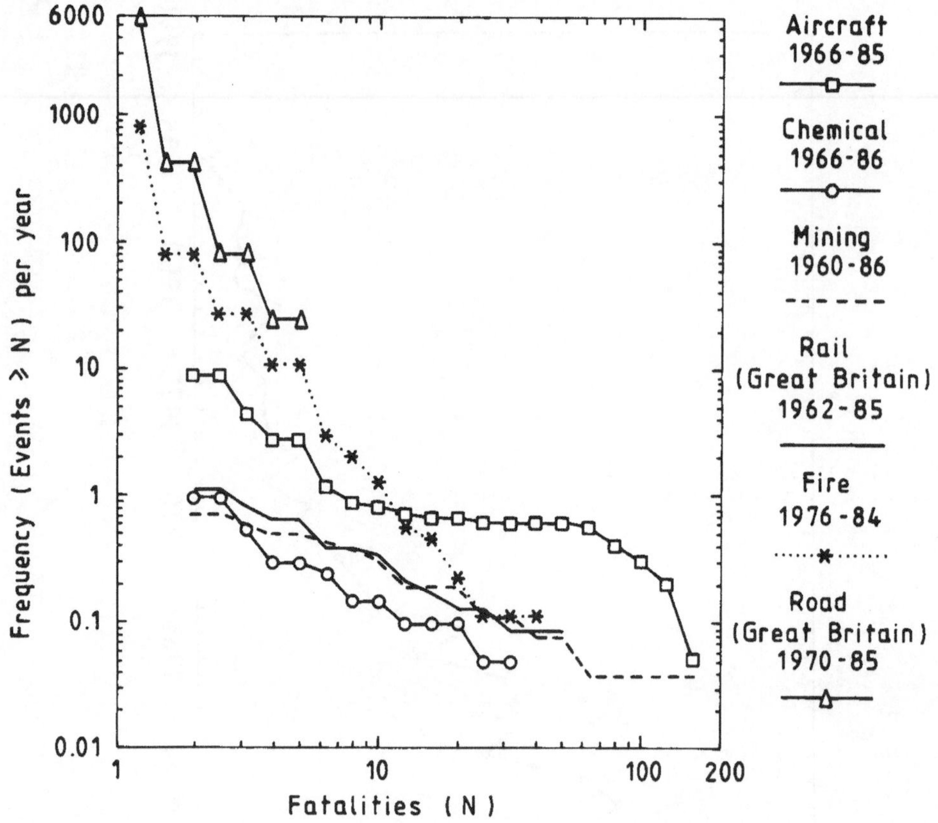

FIGURE 4.4
Frequency of Multiple Fatality Aircraft Accidents Worldwide 1946-65 and 1966-85
(Source: Fernandes-Russell, 1987b)

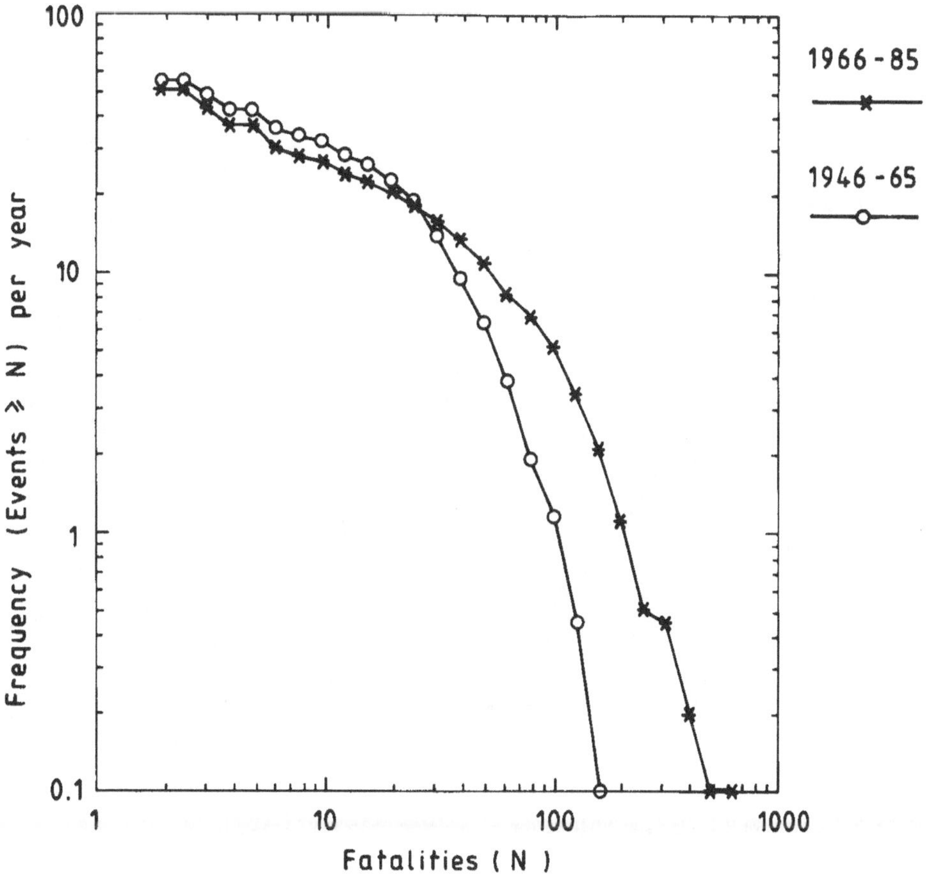

FIGURE 4.5
Frequency of Accidents in the Chemical and Petrochemical Industries UK and Worldwide 1966-86
(Source: Fernandes-Russell, 1987b)

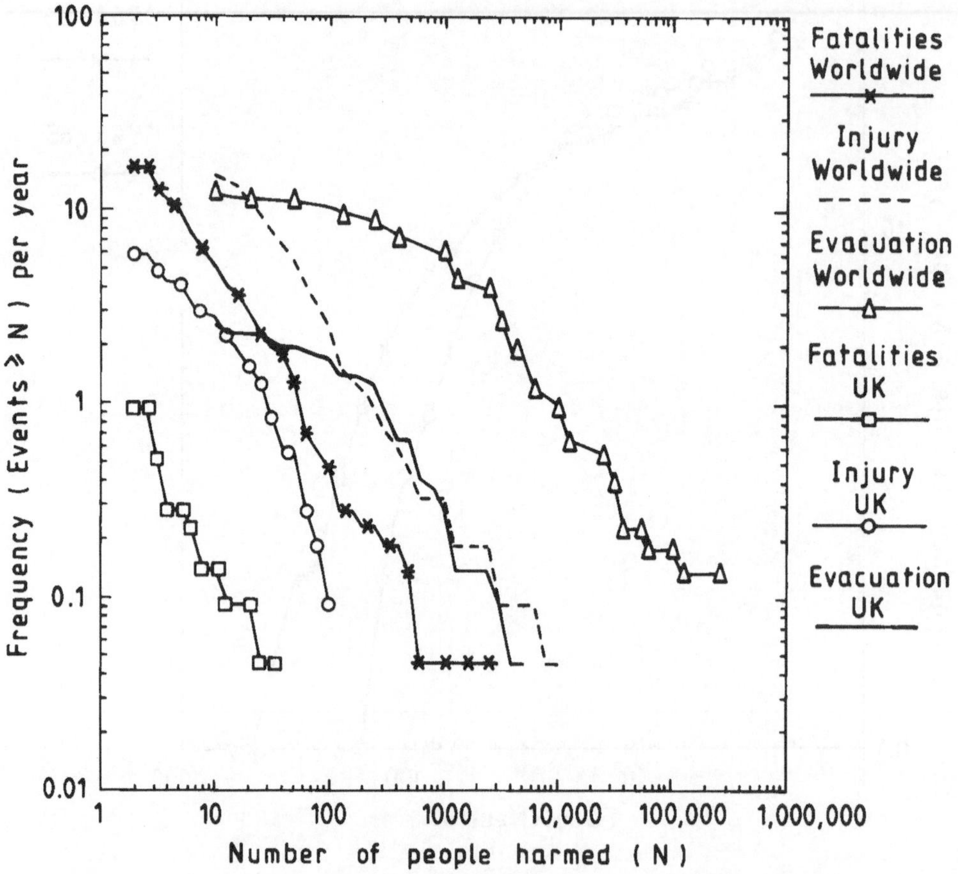

FIGURE 4.6
Frequency of Aircraft Accidents Causing Injury and Death
UK and Worldwide
(Source: Fernandes-Russell, 1987b)

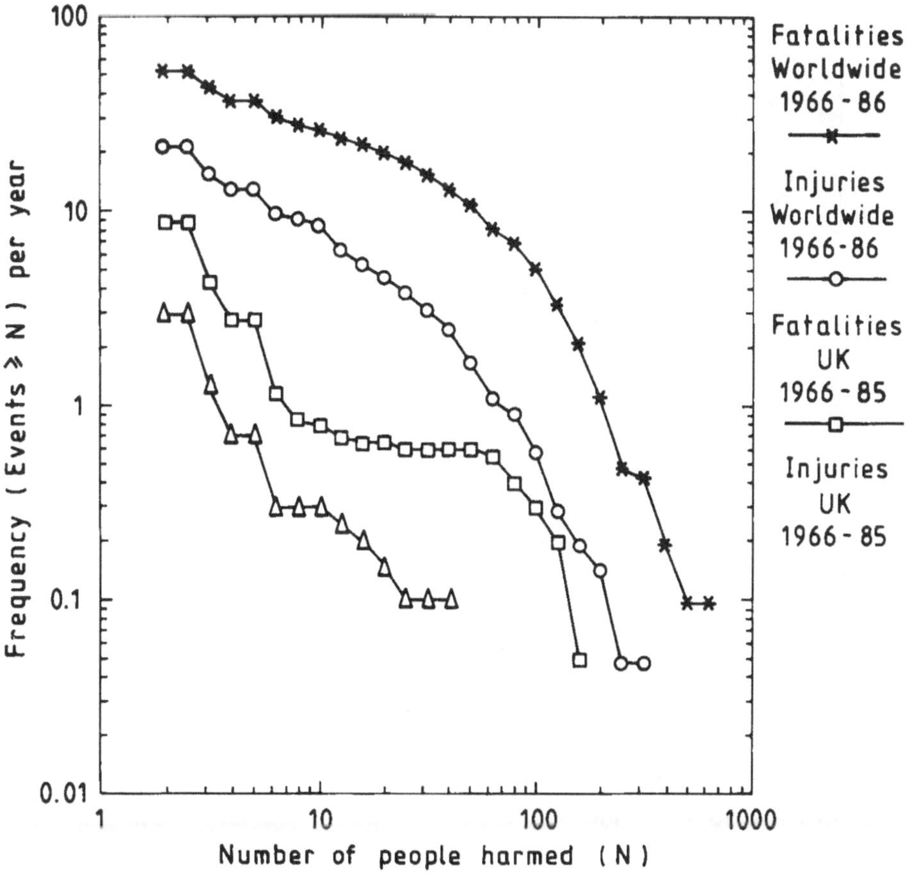

FIGURE 4.7
Frequency of Accidents Causing Fatalities
Selected Categories, Worldwide
(Source: Fernandes-Russell, 1987b)

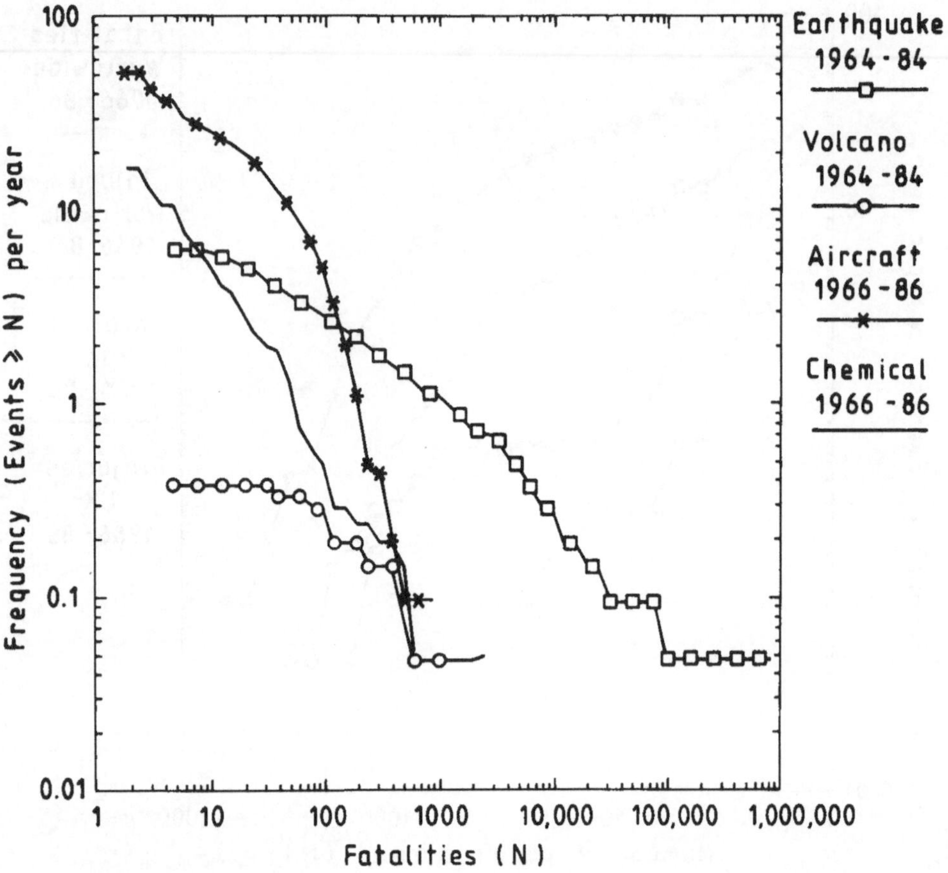

FIGURE 4.8
Frequency of Accidents Causing Injury
Selected Categories, Worldwide
(Source: Fernandes-Russell, 1987b)

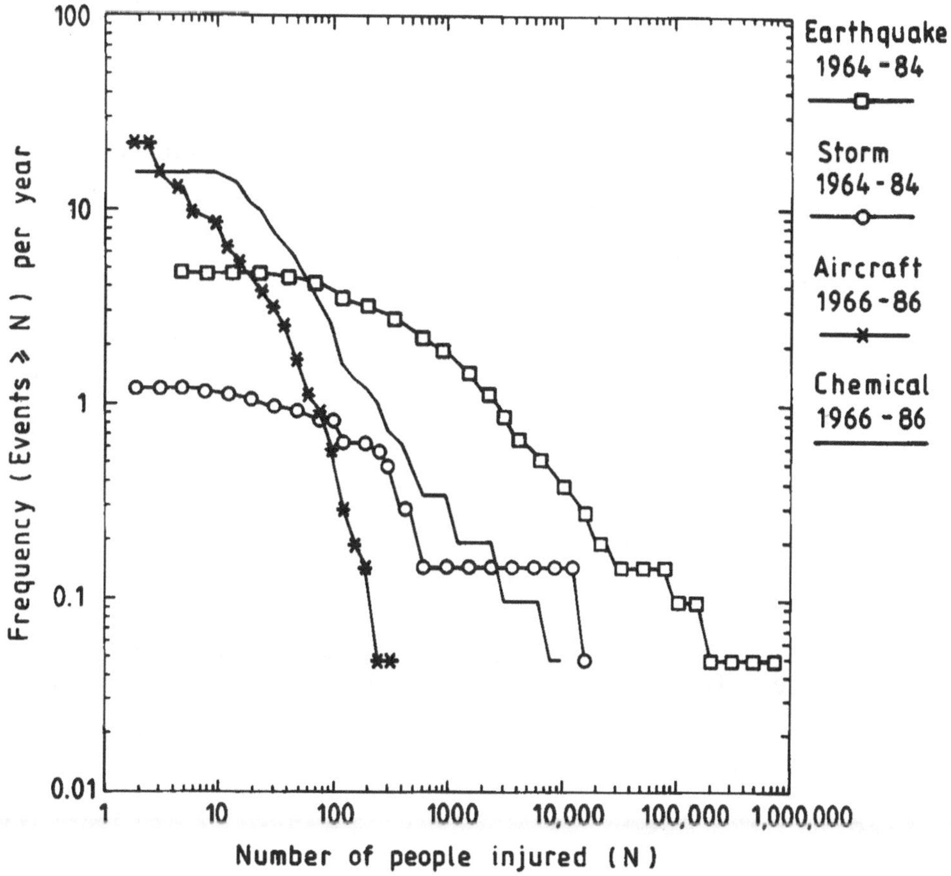

FIGURE 4.9
Comparison of Wash-1400 Results with Natural Hazards in US
(Source: Coppola and Hall, 1981)

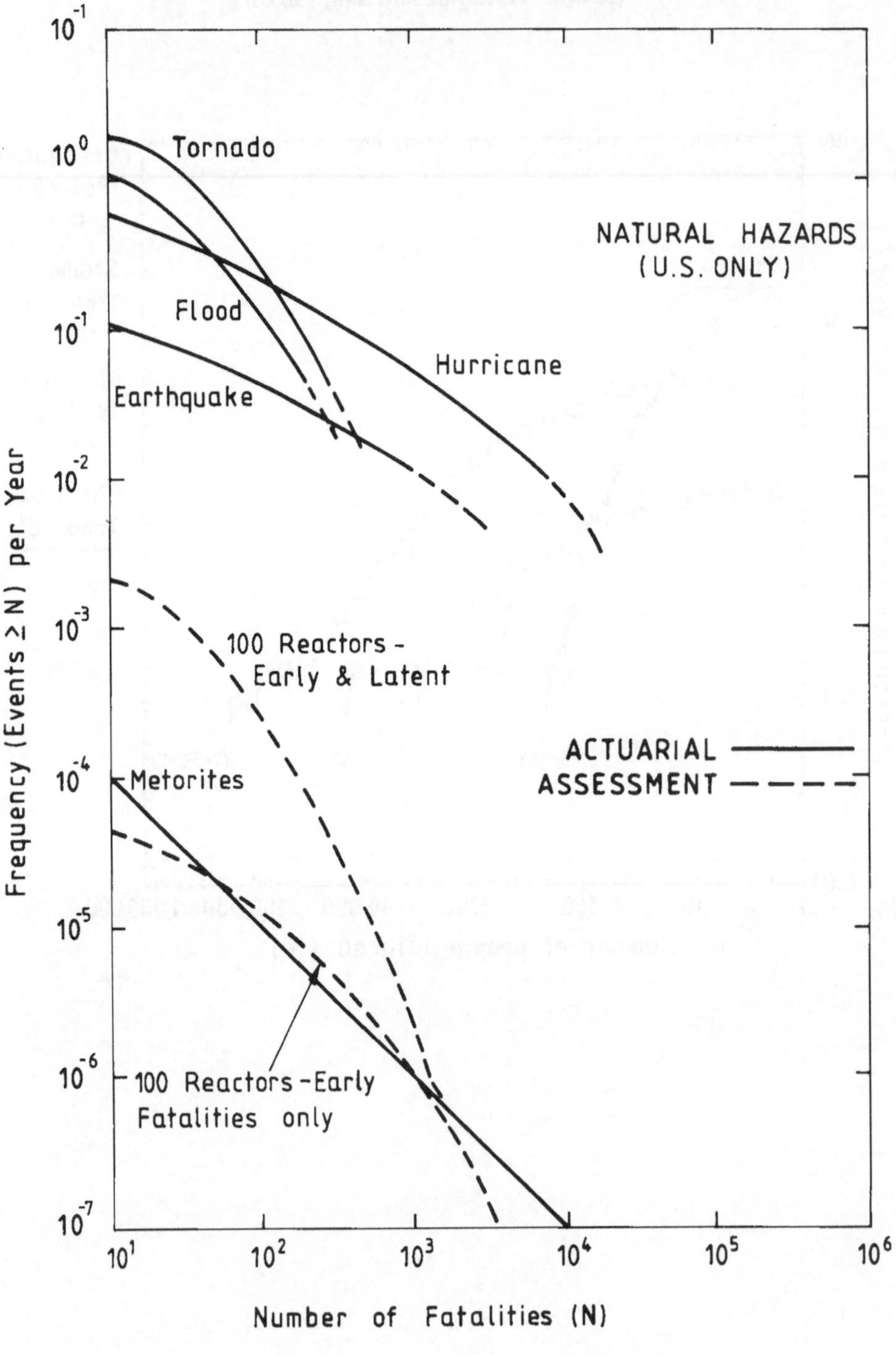

FIGURE 4.10
Comparison of Wash-1400 Results with Man-Made Hazards in US
(Source: Coppola and Hall, 1981)

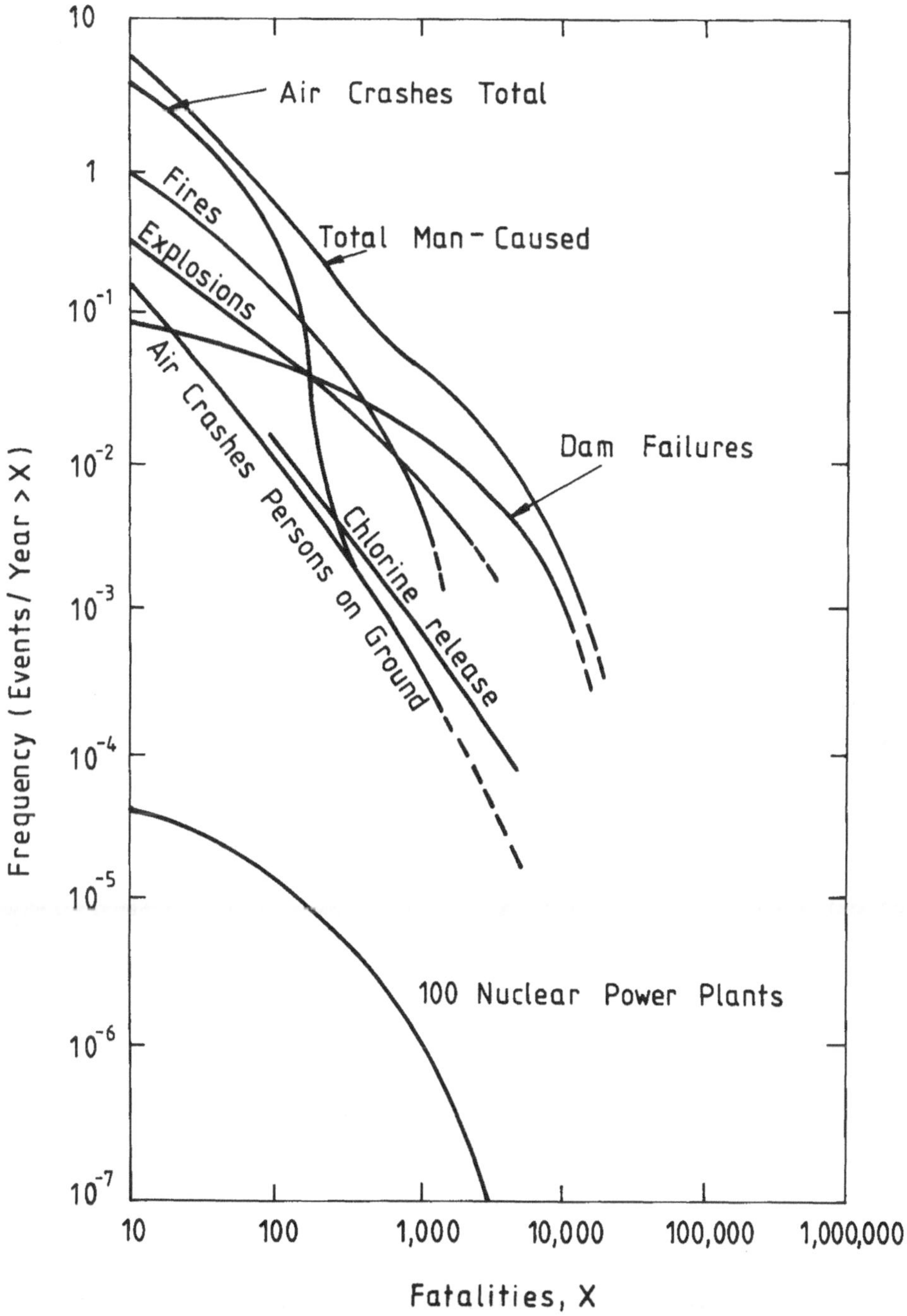

FIGURE 4.11
CCDFs for Canvey Island and Sizewell 'B'
(Adapted from Kelly & Clarke, 1982, and Barrell et al, 1985)

Note. In the case of Canvey, 'casualties' refers to serious injury or worse.

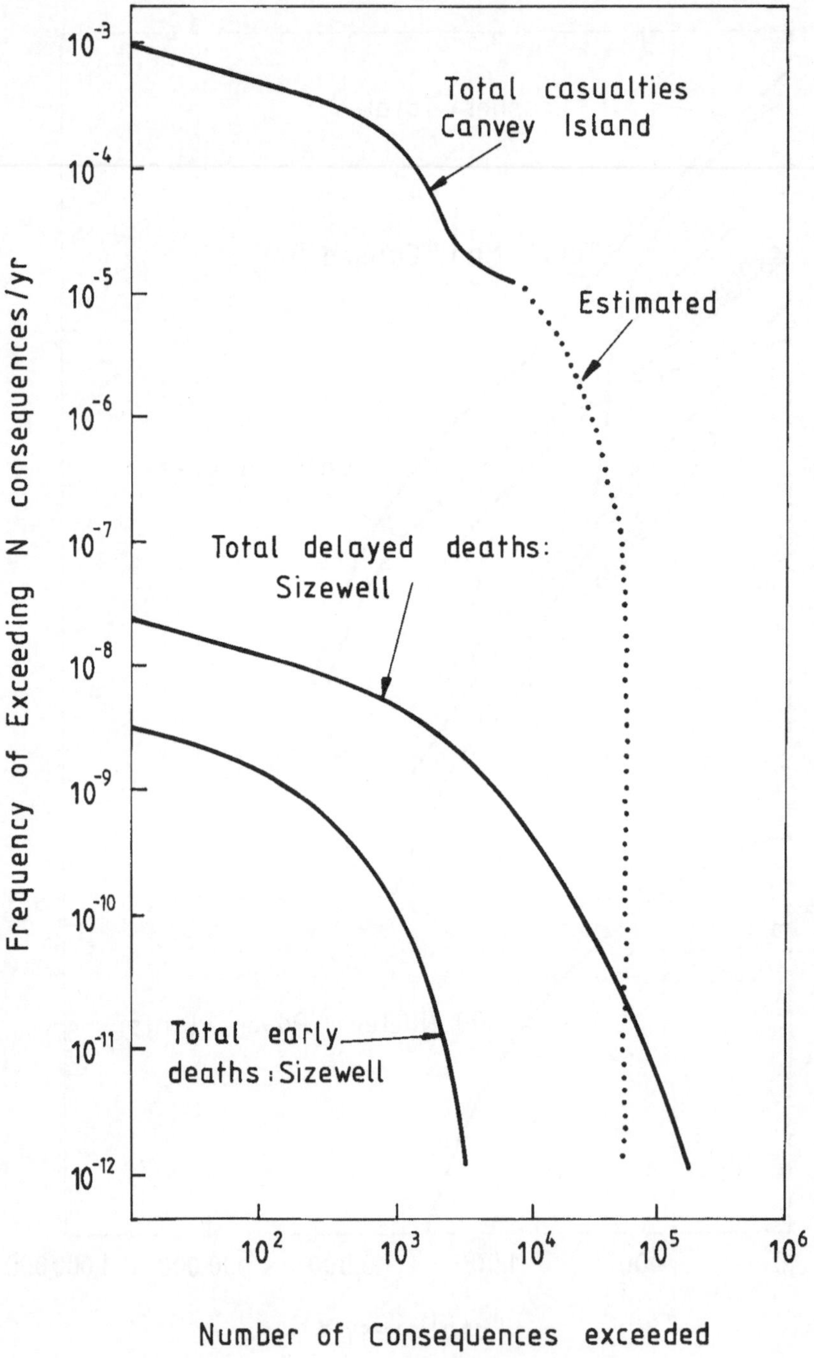

CHAPTER 5

ESTIMATING RISK: THE ANALYTICAL APPROACH

5.1 Introduction

The purpose of this chapter is to describe in general terms how risks are estimated, what problems exist in doing this, how judgement enters the process and what the uncertainties are.

The Royal Society (1983) Study Group claimed that, "risk estimation is a synthesis of science and engineering knowledge that has been brought about because it has practical utility, and not as an attempt to establish a new branch of science." This claim is reasonable, but does not, in itself, provide a case for comparing risk estimates with levels of risk thought to be tolerable. On the contrary, the techniques have been subjected to a great deal of criticism which might suggest that this should not be done. For example, Speed (1985) argues firstly that, "there is no reason at all to accept the final probability figures [of the Westinghouse Sizewell 'B' PSA] as having any value," and secondly that even, "with careful competent statistical analysis this approach could [not] yield believable probability figures." In this situation it is important to establish what models and assumptions are used in the estimation process and what can be claimed about the resulting estimates. Only then can the risk assessor judge how to use the risk estimates in reaching a properly informed decision.

We shall refer to the method to be used as the analytical approach; it is intended to be as objective and scientific as possible. Since we have defined risk in terms of probability in Chapter 2, the basic requirement is to be able to estimate probabilities. The term used for this quantitative process is probabilistic safety assessment (PSA). We shall be largely concerned, then, with the theory and practice of PSA.

A simple example may clarify the purpose of the chapter. Consider an experimental programme which determines that the half-life of a particular radionuclide to a particular decay process is 6.93×10^5 years with some stated error. Then consider an estimate of the probability of an uncontrolled release from a nuclear reactor of 10^{-6} per year. The numbers are equivalent, but the risk evaluator will form a different view of the significance of the first number compared with the second. The reason for this is that the half-life assessment can be carried out using well understood, objective scientific techniques, whereas the risk estimate inevitably involves a large amount of 'engineering judgement' or 'subjectivity.' Indeed, the first judgement is that the (implied) statistical model is appropriate. The essence of the analytical approach is to minimise the subjective content of a risk estimate, but it remains important to examine how judgements enter the process, and evaluate the estimates in the light of this.

The remainder of the chapter is organised as follows. Section 5.2 discusses the basic concepts involved in PSA while section 5.3 describes how the method is implemented for nuclear reactors. This is not strictly necessary for the present purpose, but it does provide useful background. Section 5.4 lists some of the problems which exist with the PSA technique; it is the judgements taken to resolve these problems that have to be evaluated by the risk assessor.

5.2 Principles of PSA

The purpose of this section is to give an overview of the ideas and techniques underlying PSA. In subsection 5.2.1 a brief description of the method is given which will be amplified in section 5.3 for the case of nuclear reactors. Subsection 5.2.2 contrasts PSA with what could be achieved without probabilities in order to provide further clarification of the analytical method.

5.2.1 Basic ideas

The fundamental objective of PSA is the estimation of probabilities for events of safety concern. This has already been done in Chapter 4 which presents direct estimates of risk levels, that is, the number of events is divided by the total experience. The difference between the analytical approach and the procedure of Chapter 4 springs from the inadequate database for direct calculation. This leads to the need to construct a model of the system of interest which enables data relevant to subsytems or components to be acquired and used to estimate the important probabilities. However, the need to check consistency of the results of PSA with those calculated directly should always be borne in mind (see, for example, the US Precursor Study (USNRC, 1982)).

PSA includes two types of analysis: a logical examination of the system; and physical models of the progression of accidents. The use of systems analytical models has its roots in reliability theory which was developed in the aircraft and electronics industries from the 1940s onwards. The logic models used can give qualitative insights even before they are quantified. However, the main concern here is with numerical results and these are obtained using input data much like that employed in Chapter 4: failure probabilities estimated from observed events data in component or system populations. The probabilities of other events are then calculated from the logic structures.

Now this part of the analysis may tell us the probability with which the plant will get into a dangerous condition, but it is not sufficient to tell us the probabilities of the events of true safety significance. For example, in the case of individual risk the event might be, "death due to radiation of the person most exposed," and for social risk it might be, "an accident causing *N* or more fatal cancers world-wide" (for some range of *N*). In order to calculate the probability of these events we have to use physical models to estimate the consequences of plant damage. (The jargon for this is phenomenological models, to make clear that chemical, biological and psychological effects are considered.) Here, the data are very different, and consist of experimental results, which are frequently not directly relevant, and theoretical calculations. The latter are, for the most part, computerised, and very complex.

It is generally accepted that in order to be credible a PSA should provide some estimate of the uncertainty in its results. It is clear that the uncertainty is invariably large. For the systems analysis stage there are fairly good statistical techniques for revealing the uncertainty due to sparse failure data, though problem areas remain.

The nature of the uncertainty in the physical modelling is rather different. It is necessary to extrapolate the results of experiments to the situation of interest; even for well studied problems the validity of the computer codes may remain in doubt; and for less studied phenomena the models may be very crude. In all these cases it is necessary for expert judgement to be used to quantify the PSA and the uncertainty. This will be discussed in greater detail in section 5.4. It is probably the most prominent area where subjectivity enters PSA, but, as we shall see, there are many others.

5.2.2 Features of the PSA approach

Probabilistic methods are a relatively recent development. Alternative methods are generally known as deterministic, though this is not a well defined term, and varies in meaning between countries. The intention of this subsection is to illustrate some features of probabilistic methods by comparing them to what could be achieved without probabilities.

The following discussion indicates the importance of judgement in safety assessment, however it is carried out. This was also highlighted by Layfield (1987) who devoted a chapter (18) to the topic. He notes its ubiquity "at all stages of design and safety analysis" including accident probability analysis (synonymous with PSA). He also comments in some detail on the correct balance between "well-established engineering techniques" and probabilistic methods (47.24, 14.A, 17.E, 28.125).

We shall not address the question of the role of PSA in an overall safety case though we should point out that current and forseeable practice will be a mixture of deterministic and probabilistic methods. This can be put another way, by saying that quantitative safety goals cannot be the complete determinants of risk tolerability. The way in which each approach is used in the CEGB's safety case for Sizewell 'B' will be described shortly.

It is possible to identify several features of a non-probabilistic approach:

- setting good engineering standards for the design, manufacture, installation and maintenance of components and structures,
- judging some accidents to be 'incredible',
- judging a particular event to be the 'maximum credible accident' for the purpose of detailed consequence calculations,
- defining the set of 'design basis accidents' which the plant must be proved to withstand, or which must be proved not to give doses exceeding suitably judged maxima,
- such proofs to involve conservative engineering methods.

By contrast the PSA approach would aspire to:

- take engineering standards into account in a quantitative way,
- include all conceivable accidents up to the point where their probability can be objectively assessed to be negligible, bearing their consequences in mind, using a rational approximation procedure, and
- use 'realistic' calculational methods as well as conservative ones, and to take account of both in a quantitative representation of uncertainty which will indicate the degree of pessimism in the conservative techniques.

All the quantitative inputs proposed here would have a substantial amount of judgemental input, but the probabilistic format allows these judgements to be made on a highly structured basis. Moreover, compared with the judgements made in a non-probabilistic approach, they can be more closely related to events of true safety significance. Finally it is very important to appreciate that a large part of the impact of PSA is due to the qualitative results which come from the systematic, structured analysis methods used to approach completeness, and has nothing to do with probability estimation.

Several problems have in the past been used to criticise the PSA method (and by implication to favour non-probabilistic approaches). They are mostly discussed in section 5.4, and include uncertainty, completeness, dependent failures, human factors and external events (or hazards - see subsection 5.3.1). However, none of these problems suggest that PSA should be abandoned; indeed they were for the most part highlighted by the application of PSA techniques, not created by them.

It is interesting to consider the way in which deterministic and probabilistic methods are used in the UK. The approach to analysing faults which is taken by Nuclear Electric in licensing reactors is split into a 'design basis' analysis and a 'beyond design basis' analysis. The design basis analysis corresponds to the selection of certain faults which the plant must be designed to withstand as described under deterministic methods above. This analysis uses prescribed, conservative methods. However, these faults are categorised in terms of frequency, and these categories are used to define the degree of redundancy and diversity which is required. These are probabilistically motivated concepts. In addition there are further probabilistic conditions on radioactive releases which have to be met by these 'design basis faults'. Thus the design basis analysis in the UK is considerably more probabilistic in nature than in other countries where it is truly deterministic. By contrast the treatment of the most serious accident sequences in the beyond design basis analysis is purely probabilistic. In fact for licensing Sizewell 'B' the CEGB simply calculated the frequency of 'conditions outside the design basis' in the expectation that this would be sufficiently low, so avoiding the need for further calculation of the consequences of the various minor and major accidents this encompasses.

In the light of this discussion we would therefore regard the analytical approach as incorporating the structured, systems modelling methods of PSA together with the appropriate quantification techniques, taking uncertainty into account. This will inevitably involve a great deal of judgemental input, which will be described in more detail in subsequent sections, but these judgements are made in the light of, and co-exist with, the maximum amount of objective, and recorded, information (including that arising from deterministic analysis). It is the sole means of estimating risk.

However, it must be borne in mind that if there is some aspect of plant safety which cannot be treated in this way, the analytical method will have to adopt the kind of judgement characteristic of the 'deterministic' approach (see subsection 5.4.3).

5.3 PSA for Nuclear Reactors

Although the idea of defining a complete set of events and associating probabilities with them is a simple one, in practice its application to a complex industrial plant is far from straightforward. As stated in section 5.1, the difficulties which are involved will need to be appreciated by the risk evaluator who wishes to use the results of an assessment. These difficulties will be outlined in section 5.4, but, in order to provide some background, this section describes the various stages of a PSA with particular reference to nuclear reactors. Chapter 7 discusses the historical development of the technique and its use in regulation in both the nuclear and non-nuclear contexts.

Although we shall describe the particular techniques used when carrying out a probabilistic assessment of nuclear reactor safety, it will be appreciated that a very similar process is carried out for a chemical plant, for example. In many cases, such as systems analysis, the methods will be the same, in others, such as quantifying the health effects of a particular chemical, they are merely analogous.

The process which will be described is complex and involves many physical models and logic structures. Its application involves a large number of judgements and approximations. Not all the approximations are recognised as such, and tend to be treated as part of the method. One class of approximation is those made to provide a very coarse risk estimate, which is, nonetheless, adequate to ensure compliance with appropriate targets. This is an important area of development within PSA which should ensure better matching of effort to safety significance. Past experience, and the description below, suggest that PSAs are large, complicated and difficult to follow; for simple, safe plant, this should not be the case if only a bounding estimate is required.

If a PSA is viewed in terms of estimating frequencies, then four major classes of event can be identified, and, correspondingly, three types of analysis are necessary to move from one class to the next:

Event Class	Type of Analysis	PSA Level
Initiating Events		
	Plant Analysis	
Plant Damage States		1
	Containment Analysis and Fission Product Transport	
Source Terms		2
	Consequence Calculations	
Consequences		3

Depending on the stage of analysis which is reached, a PSA is designated as Level 1, 2 or 3, as indicated in the right hand column (USNRC, 1983). This forms a useful classification for defining the event classes and describing the required analysis. This is done in the succeeding subsections.

5.3.1 Plant analysis

This begins with the various initiating events which have the potential to lead to undesired consequences and analyses the first stages of their development. Its scope is defined by those regimes in which the physical behaviour of the reactor and its fuel is well modelled, and determined by the operation of the reactor systems. These physical models are used in the design of the reactor for transient analysis (that is, simulating the behaviour of the reactor under time dependent operational and accident conditions) and when the accident has escalated such that the models are no longer valid, generally when core geometry is first lost, conditions outside the design basis are said to have been reached. Thus a Level 1 PSA can be carried out using physical models which already exist, and by analysing the reliability of the various reactor systems.

Plant analysis will begin by defining the initiating events of interest. These are of two kinds: internal plant faults and hazards. Internal plant faults are initiators which result from identified failures or transients within the reactor systems whereas hazards are of more general causation. (Note that the term 'hazard' is used in a different sense in this section than elsewhere in this report. Hazards are also known as external events, but this is confusing for internal hazards.) Internal hazards include fire, explosion and missiles generated by turbine disintegration and originate within the plant. External hazards include earthquake, aircraft crash and extreme weather. Because hazards have the capacity to cause multiple failures, they cause special problems which often lead to them being the subject of a separate analysis.

The next stage in a Level 1 PSA will generally be to define the plant damage states which are to be used. This will depend on the purpose of the PSA. If it is to stop at Level 1, for example to assess the design engineering of the reactor, a single state such as 'core melt' or 'conditions outside the design basis' often suffices. If the PSA is to be extended to Level 2 or 3, then a whole spectrum of plant damage states, such as mode of core melt, is necessary to provide suitably precise initial conditions for the next stage. The first guess at a suitable set of plant damage states may have to be revised in the light of further analysis. The initiating faults and the plant damage states are generally linked by means of an event tree representing the success or failure of the various reactor systems.

Fault and event trees are the two predominant types of logic structure used in PSA. Simple examples are shown in Figures 5.1 and 5.2. Although they are formally equivalent, the basic difference in practice is that event trees are used to analyse the various outcomes stemming from an event, whereas fault trees look at the combinations of events leading to a defined event. Thus an event tree can represent the various accident sequences resulting from a given initiator and can be used to identify the plant damage state resulting from each sequence. A fault tree can be used to identify the component failures which lead to a particular system failure which then determines the outcome of a logic node on an event tree. However, the precise way in which these logic trees are used is very dependent on the scope and purpose of the analysis and over-generalisations should be avoided.

Prior to drawing the fault and event trees, schedules of system requirements are drawn up, and preliminary failure analyses such as Hazard and Operability Studies (HAZOPS) or Failure Mode and Effect Analysis (FMEA) are carried out as an aid to completeness in constructing the trees. These are described in the PRA Procedures Guide (USNRC, 1983). The system requirements (or 'success criteria') have to be determined for each initiating event by using the design codes for transient analysis. In general these codes will give accurate simulations of plant behaviour, but for some accident conditions, for example loss of coolant in PWRs, they are intended to be conservative.

None of the work described so far involves the calculation of frequencies, which is the main aim of a PSA. However, useful engineering insights are often provided by carrying out a systematic, structured investigation of the reactor systems. For example it may be possible to identify ways in which failures of single components can cause accident escalation because of overlooked dependencies in the systems.

In order to determine the frequencies of the plant damage states, it is necessary to quantify the trees. The requirements for this are the frequencies of the initiating faults and the probabilities of the various component failures on the fault trees. Both of these are obtained from failure data which is stored in computer databanks, or, in the case of an operating plant, from experience for that plant. When suitable assumptions about testing, maintenance and repair have been made, computer programs can analyse the logic trees to give the plant damage state frequencies.

5.3.2 Containment analysis and fission product transport

This stage of a PSA considers how the accident sequences resulting in plant damage states such as core melt can evolve to give a radioactive release or source term. This takes the PSA from Level 1 to Level 2. Its defining feature is that the sequences are predominantly a series of physical processes which evolve largely unaffected by the reactor systems, though some containment systems may be modelled in this phase. The models for these physical processes are a good deal more uncertain than those used in transient analysis.

Ideally one should begin by defining a suitable set of source terms. These specify not only the amounts of the radioactive materials released, but also the temperature, height and duration of the release. They are linked back to the plant damage states by the containment event tree. The nodes of the event tree are selected in such a way that the physical processes, and operation of the containment systems, are represented with a suitable degree of definition so that the source term appropriate to the end point of each path can be identified.

Containment analysis then involves identifying the path or paths which will be taken through the containment event tree given some plant damage state, and fission product transport calculates the source term consequent on each path. The two are not entirely independent, since the location and form of the fission products can affect the containment behaviour.

The physical processes which can enter the containment analysis are many. They include: release of fission products from the fuel matrix and from the fuel element; their transport in the reactor coolant system and in the containment; the failure of structures such as the reactor vessel and the reactor building; the processes leading to stressing these structures such as steam explosions, fission product heating, coolant flashing to steam, hydrogen explosions, sodium fires and core/coolant interactions with concrete; and the effectiveness of the containment systems such as coolers, suppression pools, recombiners and filters.

It will be appreciated that many of these effects are very crudely modelled, and the containment event tree is normally used to represent the uncertainty attached to this by assigning probabilities on the tree appropriately. To understand this one can start from a picture in which a plant damage state is sufficiently well defined that a single defined release occurs. The physical processes evolve deterministically. This ignores failures in the containment systems which will be random. However, by inserting probabilities on the event tree so that the sequence is not prescribed at each node a number of the source terms will result from a single plant damage state. This could happen because each plant damage state represents a range of possible conditions, or because a physical process is genuinely random, but this is not the case in general and probabilities on the containment event tree represent uncertainty due to inadequacies in the physical modelling. We shall return to this point in section 5.4.

When the nodes on the containment event tree have been quantified (for each plant damage state) the frequency of each release can be obtained by computer analysis.

Because source terms are multi-dimensional quantities (releases for each radionuclide, temperature, height, time, duration), it is rather difficult to pick a suitable representative set with a manageable number of members in the light of the containment analysis, let alone before it (which is desirable if it is intended to pursue each stage of the PSA simultaneously). However, this technique is an illustration of one of the most important approximating methods in PSA, that of categorisation or binning, in which a representative set is selected to cover all possibilities. Plant damage states are another example, and the same idea can be used to reduce the number of initiating events to be considered.

5.3.3 Consequence analysis

This moves the PSA forward from Level 2 to Level 3. It examines how the various releases can lead to undesired health or social effects and hence risk, represented in the ways described in Chapter 3.

The physical processes involved here are: dispersion of radionuclides in the atmosphere in different weather conditions; processes which lead to their deposition; their radioactive decay; their dispersion in marine and ground water environments; their take-up by man, animals and plants; their dispersion in food chains; the radiation doses received by direct shine, ingestion and inhalation routes; the health effects of these doses. Account must also be taken of countermeasures intended to reduce the consequences of an accident such as evacuation and the interdiction of land and foodstuffs. The social consequences of an accident such as the cost of countermeasures, of cleanup and of compensation must also be evaluated. All the concerns identified in Chapter 3 may be considered.

This stage does not use logic models, but instead employs large computer programs to calculate the consequences of a given release taking the randomness due to various weather conditions into account. Clearly the uncertainty in modelling some of the process is large, but this is less explicit than in the containment analysis stage where it is represented on the event tree. The other random aspect of consequence calculations is that of stochastic health effects, discussed in Chapter 3.

5.4 Problem Areas in PSA

In this section we move from the basic ideas on PSA in the two previous sections to consider what its shortcomings are. As we said in section 5.1, it is important to bear these in mind when using PSA results in a risk evaluation. In short, we shall be discussing PSA uncertainty.

Once again the discussion is mainly concerned with assessing nuclear reactors. The importance of the various factors will vary depending on the type of plant under consideration, and the targets against which the risk estimates are to be assessed. New contexts, for example waste repositories, will generally throw up new problems.

5.4.1 Problems with logic models

(i) Completeness

The concern here is whether the logic models represent all possible effects. Some causes of failure may have been omitted or some physical process ignored which contributes to or alters an accident sequence. We are really only concerned here with effects that the analyst does not think of. Any effects which are imagined, but not included explicitly in the analysis, should have their omission explained by some argument, albeit judgemental. The latter becomes an approximation rather than failure of completeness.

Although there can never be a a guarantee of completeness, it is clear that the discipline imposed by the necessity to construct logic models is a considerable help in approaching the goal of completeness. Thus, this is not viewed as a major problem, except perhaps in the human factors field discussed below.

A problem related to completeness concerns the ability of fault and event trees to represent accident sequences adequately. The functioning of systems can probably be covered by fault tree success/failure logic, although in some cases careful definitions may be needed. More importantly, the physical processes represented on event trees are often not well suited to the branching logic. For example, the pressure generated by some effect may be enough to fail a structure completely, partially or not at all. Thus a three branch node with careful definitions will be needed. However, each branch will cover a range of circumstances and, in particular, this could cause problems for the quantification of succeeding nodes.

(ii) Dependent Failures

A long-standing difficulty in the plant analysis stage of a PSA is where there is a causal link between two or more failures which means that they cannot be considered independent. In particular, systems which achieve high reliability by using redundant trains of identical equipment are prone to the special case of common mode failure; for example, all the pumps might fail due to a repeated maintenance error.

Many causes of dependent failure can be identified and represented on the logic models. However, it is not possible to cover all causes, and this has resulted in techniques to cover unidentified dependences. These models, and their quantification from failure data, are described in Fleming et al (1986), for example. The situation is that good models are becoming available, but they are still not routinely integrated into PSAs. The treatment of dependent failure in any given assessment must, therefore, be very carefully examined.

(iii) Databases

There are standard statistical techniques for using failure data to quantify fault trees built up from component failure probabilities. These techniques are based on the assumption that a series of components identical to the plant item has been tested, and the failures noted. In practice, the data comes from operating plant which is rarely identical to that being assessed and which is stored in databanks. Even if there is plant specific data available, there is unlikely to be much relevant experience, especially since any persistent cause of failure is likely to be eliminated. A major question is thus that of data applicability, and judgement must be exercised to obtain a set of data which gives useful results, without stretching its applicability too much. This is further discussed by Mosleh (1986).

5.4.2 Problems with physical models

The problems with physical models arise not in the plant analysis stage where the physical behaviour is predicted by fairly well validated transient codes, but in the subsequent containment analysis, fission product transport and consequence calculation stages. The physical processes which take place during accident sequences are not as well understood for the reason that there is very little data from accidents themselves, and the evolution has to be modelled by separating out the various phenomena and performing experiments, often at reduced scale, to determine what occurs. These are extrapolated back to accident conditions with the use of complex computer codes, whose validation is very difficult.

Of the three stages mentioned above, consequence calculation is the least uncertain since processes such as atmospheric dispersion, deposition and radionuclide take-up are fairly well understood. The problems here are to model the statistical distribution of weather conditions adequately, especially those rare conditions causing very high consequences, and to maintain realism out to large distances.

There is an important difference in the approach to uncertainty between containment analysis and fission product transport. The response of the containment to accident conditions is modelled using an event tree whereas fission product transport, like consequence calculation, is modelled by deterministic computer codes. That is, the codes produce a single result from given initial conditions. The containment event tree, however, can be quantified by probabilities which can be used to represent uncertainty in the physical modelling, as well as genuine randomness. This was explained in subsection 5.3.2. Thus uncertainty is incorporated into the analysis more explicitly.

This causes a potential problem in that the representation of uncertainty may be unreasonably reduced. The reason for this is that event tree probabilities become part of the overall statistical model which assigns probabilities to events. If we wish to assess the uncertainty in these probabilities (using the methods to be described in subsection 5.4.5) we must vary the parameters involved in their calculation. Since these include the containment event tree probabilities, which to some extent already represent uncertainty, this is somewhat paradoxical. The usual solution is to vary them anyway. The practical justification is that in assigning subjective event tree probabilities, the assessor will be uncertain as to the appropriate values, and a subjective uncertainty range can represent this.

The physical processes which contribute most to the uncertainty depend on the type of reactor which is being analysed. For example, for PWRs some of the major sources of uncertainty are the generation of hydrogen during core melt, the capability of steam explosions to fail the reactor vessel and containment simultaneously, the chemical form of iodine and the strength and failure mode of the containment. For sodium cooled fast reactors, the analogous phenomena are interactions between molten fuel and coolant, recriticality and the strength of the vessel and reactor building.

5.4.3 Human factors

The interaction of humans with a nuclear or other process plant is a special cause for concern. This arises because humans do not have the predictable failure characteristics of inanimate components, and this calls into doubt the PSA method of assigning probabilities to events.

It is apparent that some types of human error, such as turning the wrong switch, can be adequately characterised by probabilities estimated from data. Such failures to execute a plan correctly are known as 'slips' (Reason, 1986) in contrast to 'mistakes' where the failure involves a wrong plan. Accidents in which mistakes were a dominating feature include Three Mile Island (mis-diagnosis) and Chernobyl (malpractice). At the present time it is not standard practice in PSA to attempt to identify mistakes, though some attempts are being made to find dominating ones, and so completeness is a problem. However, it is possible to assess the robustness of a system against mistakes, taking account of ergonomic and good design principles as well as effective operator selection, training and performance monitoring. It is important that such an assessment form part of the safety case considered although it would be only partially, if at all, probabilistic in character. It follows that here, at least, there is a contribution to 'risk' which it may not prove possible to estimate convincingly using probabilities. As a result, a 'deterministic' approach, as described in subsection 5.2.2, may be appropriate. Such an approach, considering human factors separately from probabilistic results, has been used by the NII in licensing Sizewell 'B' (Campbell, 1985; Layfield, 1987).

5.4.4 Hazards

Estimating the contribution to risk from internal and external hazards causes special difficulties. One of these is their ability to cause a large number of failures simultaneously. Not only can several of the systems be affected in the plant analysis stage, but the dependencies can also extend to other stages. For example an earthquake can cause a core melt, fail the containment and cause enough social disruption to prevent evacuation. This necessitates either adaption of the logic trees used for internal initiating events, or even the development of new trees.

Another problem is assessing the effect of very rare events. The ability of the PSA approach to estimate small frequencies of events stemming from plant faults is due to combining several probabilities, each of which can be determined reasonably accurately. It is very hard to estimate the size of a similarly rare single event, such as an earthquake, with the same credibility. Much the same difficulty applies to rare internal initiating events such as pressure vessel failure.

In some countries these difficulties have led to a semi-deterministic approach to hazard analysis. For example, even in the Netherlands where risk standards are comparatively highly developed, an entirely separate approach is used for hazards, involving requirements for external events such as floods relating to defined probabilities of occurrence.

5.4.5 Representing uncertainty quantitatively

When a PSA has produced a probabilistic model of a plant's safety characteristics, the probability of a particular event can be calculated as a function of various parameters. This subsection briefly considers the methods available to represent the uncertainty in these parameters and propagate the uncertainty to the output probabilities. Because some of the models used in a PSA cannot be quantified using standard statistical techniques, a considerable amount of subjective input is necessary to assess the uncertainty, and this has led to considerable controversy. So far no consensus has been reached on how (or indeed whether) this should be done. In this subsection we shall briefly describe the two extreme approaches, probability distributions and bounding methods, and mention some others.

We note in passing that there is a body of opinion that considers that the uncertainties involved in assessing risk are part of the risk itself so that there is no virtue in expressing this uncertainty. Instead, a single 'point estimate' or 'best estimate', representing a subjective mean, or some other parameter would be the only output of a PSA. A similar approach is often proposed when the calculational methods are thought to be conservative. We consider that these approaches do not maximise the use of objective information and are not helpful to the decision maker and recommend that all probabilistic safety assessments should contain quantitative estimates of the uncertainty associated with important outputs so far as reasonably practicable.

We also note that non-quantitative approaches to characterising uncertainty have been proposed which are complementary to the quantitative methods. These relate to the provenance of the information used in the assessment. Thus the HSE developed an attribution code in the early 1970s which depended on characteristics such as whether numbers were derived from historical data or from subjective guesses. More recently Funtowicz and Ravetz (1987) have proposed a system based on the 'pedigree' of information. This covers attributes such as the nature of the theoretical model employed, the type of data it uses and the peer concensus enjoyed by the analysis.

The approach to uncertainty which has been most used to date is that of using subjective probability distributions. This is based on Bayesian techniques for statistical inference (Silvey, 1975). In situations where statistical data are available (for example, component failure probabilities), the distribution for each parameter can be obtained by updating using methods with a relatively small subjective content. In other situations the distribution has to be obtained by expert assessment in the light of the experimental data and calculational results. Often this is done by combining the opinions of a panel of experts.

The attractions of the method are: the probability distributions for the input parameters are easily propagated to give the distribution of the output variables using well known probability algebra; point values in terms of means or medians are easily provided; confidence statements can be made; and, finally, subjective probability provides the appropriate format for the accepted means of quantitative decision making (see Chapter 8). These reasons explain why the majority of uncertainty analyses in PSA have used subjective probability techniques.

However, there is a body of opinion which is uneasy with this. The main reasons include:

- it is inappropriate to consider uncertainty in safety in terms of betting odds (which is effectively what subjective probabilities amount to),
- the subjective probability distributions are meaningless, contain an insupportable degree of detail, do not properly represent ignorance and cannot be related to the evidence in a rigorous way, and
- the probabilistic propagation mechanism has undesirable effects, in particular it unreasonably narrows the range of possible values.

The main formal alternative is to consider a set of pessimistic values for each parameter and combine these to obtain a pessimistic output value. The same can be done with optimistic values. This is known as a bounding analysis. However, the appropriate degree of pessimism (or optimism) is a subjective judgement, and one to which the output bound is very sensitive. Thus the results should be treated with caution.

A number of other representation and propagation ideas have been investigated (Apostolakis, 1987). These methods may all be of use to the decision maker; none of them is universally accepted and none escapes criticism.

This indicates that no single method is appropriate, but that they should all be retained, so far as practicable, to give the decision maker and plant manager the greatest possible amount of information. It is important that they understand the advantages and drawbacks of each approach. Decision making is discussed in Chapter 8.

5.5 Conclusions and Recommendation

(1) We recognise that probabilistic methods, associated with quantitative safety goals, cannot provide the complete safety case for hazardous installations.

(2) In order to incorporate PSA results into a risk evaluation, account must be taken of all the subjective factors which enter, including the basic choice of statistical models. The risk evaluator must form a view on the extent to which he will accept these judgements, or what allowance to make for them in reaching decisions. In practice, of course, he will have considerable influence over the methods used and judgements made.

(3) The use of judgement in PSA has been highlighted in the following areas:
- development of fault and event trees,
- potential lack of completeness,
- selection of plant damage states,
- selection of source terms,
- selection of applicable data,
- modelling dependent failures,
- modelling of human factors,
- modelling of physical processes,
- inclusion and modelling of hazards,
- methods for representing and propagating uncertainty.

(4) There may be a need to adopt a deterministic approach in evaluating the contribution to risk due to human factors such as mis-diagnosis and malpractice.

(5) We recommend that probabilistic safety assessments should contain quantitative estimates of the uncertainty associated with important outputs so far as reasonably practicable. This does not apply where a simple analysis shows clearly that targets are met.

5.6 References

Apostolakis, G, 1987, **Uncertainty in PSA**. Paper M18/1, 9th International Conference on Structural Mechanics in Reactor Technology, Lausanne, Switzerland, 17-21 August, 1987.

Campbell, J F, 1985, **The Role of PSA in the Licensing of Sizewell 'B'**. IAEA Seminar on Implications of PRA, Blackpool, 18-22 March 1985, IAEA SR-111/31.

Fleming, K N, Mosleh, A, and Deremer, R K, 1986, **A Systematic Procedure for the Incorporation of Common Cause Events into Risk and Reliability Models**. Nuclear Engineering and Design, **93**(1&2), p 245. (FMD)

Funtowicz, S and Ravetz, J, 1987, **The Arithmetic of Scientific Uncertainty**. Phys Bull, 38, No 11, pp 412-414.

Layfield, F, 1987, **Sizewell 'B' Public Inquiry Report**. HMSO, London.

Mosleh, A, 1986, **Hidden Sources of Uncertainty: Judgement in the Collection and Analysis of Data**. Nuclear Engineering and Design, **93**, p 187.

Reason, J, 1986, **Recurrent Errors in Process Environments: Some Implications for the Design of Intelligent Decision Support Systems**. NATO Advanced Study Institute on Intelligent Decision Support in Process Environments (E Hollnagel et al, eds), Springer-Verlag.

Royal Society, 1983, **Risk Assessment: A Study Group Report**.

Silvey, S D, 1975, **Statistical Inference**. Chapman and Hall, London.

Speed, T P, 1985, **PRA in the Nuclear Industry: Wash-1400 and Beyond.** Proceedings of the Berkeley Conference in Honor of J Neyman and J Kiefer (L M Le Cam and R A Olshen, eds), Wadsworth.

USNRC, 1982, **Precursors to Severe Core Damage Accidents: 1969-1979.** NUREG/CR-2497.

USNRC, 1983, **PRA Procedures Guide.** NUREG/CR-2300.

FIGURE 5.1
Illustrative Fault Tree

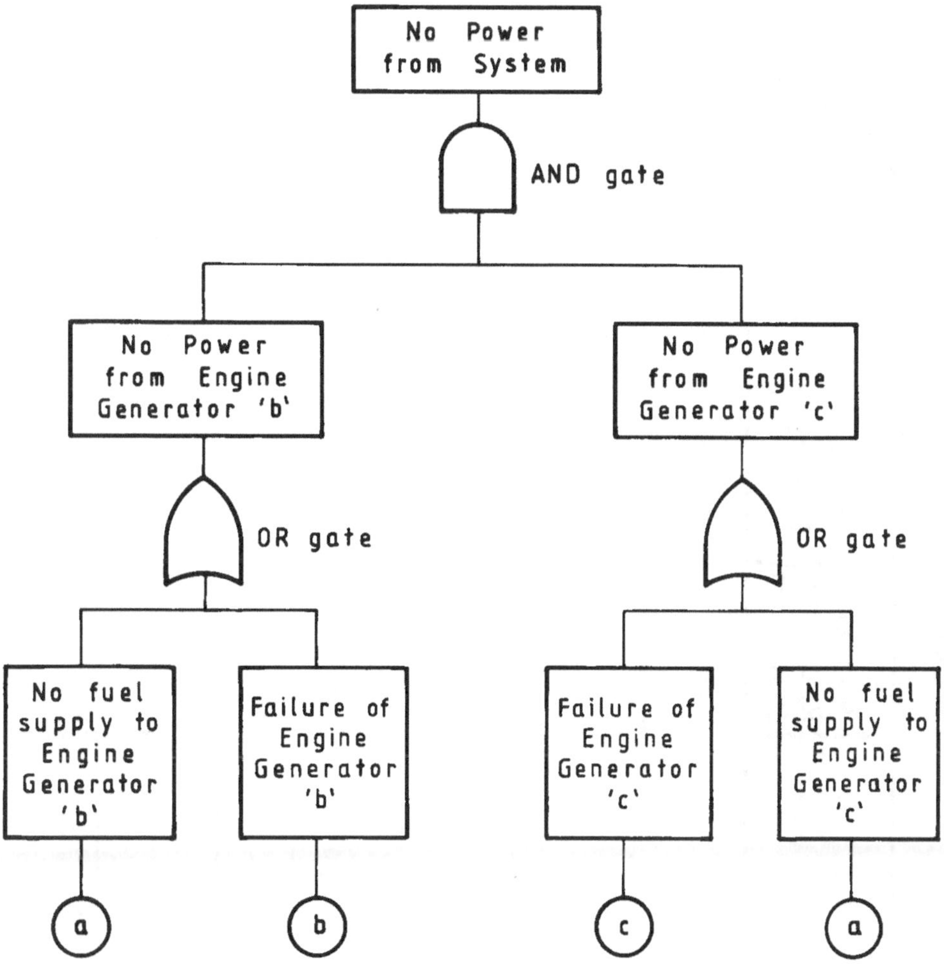

FIGURE 5.2
Illustrative Event Tree

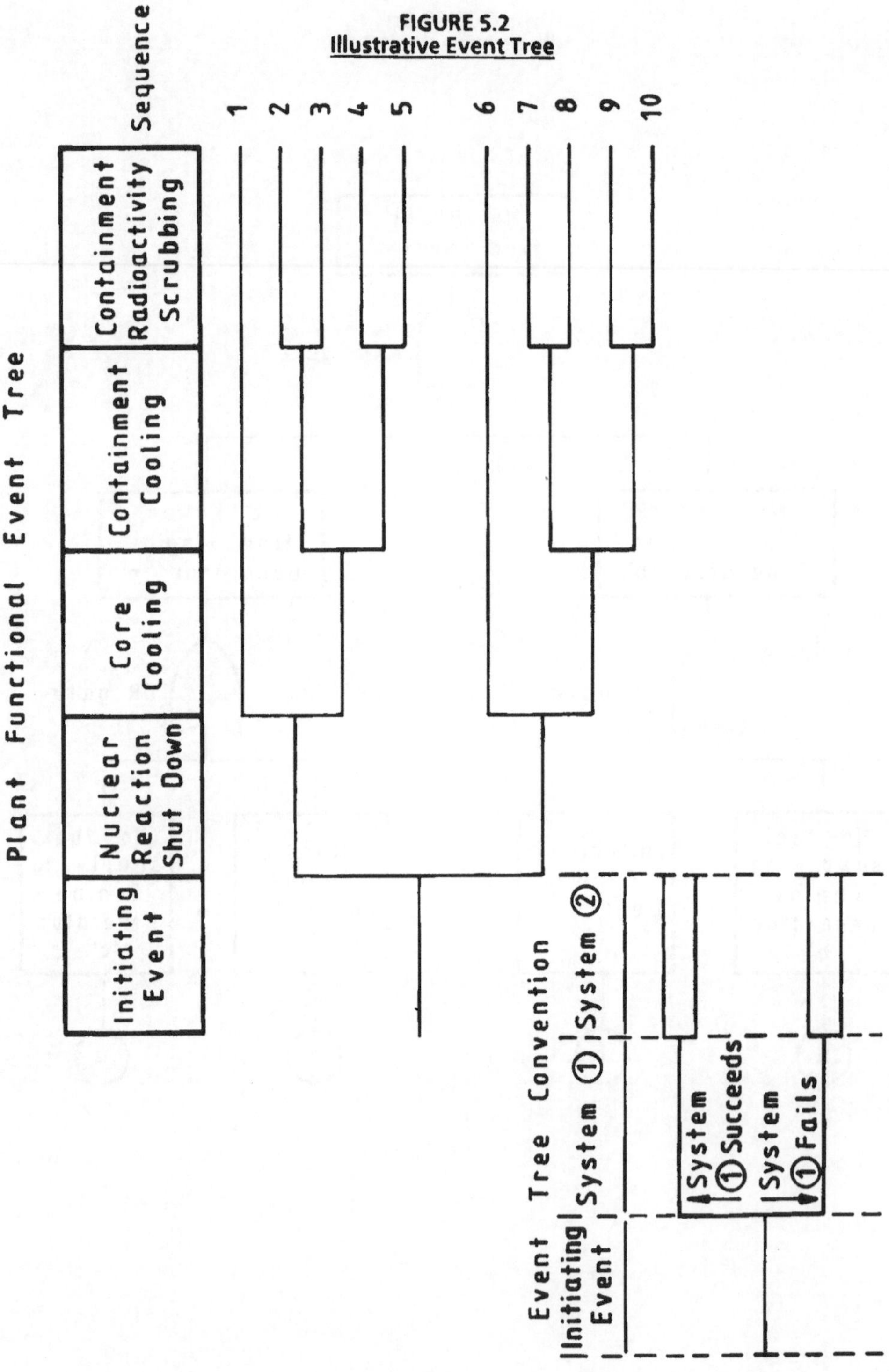

CHAPTER 6

STANDARDS FOR SOCIAL RISK EVALUATION

6.1 Introduction

This chapter covers a great deal of ground; since each section has a comprehensive introduction, the overview given here is fairly brief.

The chapter begins with a discussion of the circumstances in which risks might be tolerable in section 6.2. This can happen either if they are of no concern or if they are outweighed by the benefits. Section 6.3 discusses both of these possibilities in more detail. Particularly in the former case, but also in the latter to some extent, it is useful for the risk evaluator to be able to compare the predicted risks of hazardous installations with observed or predicted levels of existing risk. Thus section 6.4 examines what risks should be compared and sets out some principles for making such comparisons. These comparisons are carried out in sections 6.5 and 6.6 which derive reference risk levels which should be of use to the risk evaluator. Section 6.5 discusses existing risks which are, in some sense, of no concern, whereas section 6.6 considers how such levels may be derived from risks for which this is not true.

Although numerical values of some of the reference levels are set out in these two sections, they are of subsidiary importance to the description of the principles by which they may be obtained.

6.2 Risk Acceptance and Risk Evaluation

> ACCEPT: to take (something offered): to receive (with approbation, favour, consent, resignation or passivity): ...

The dictionary definition reflects the many shades of attitudes associated with "acceptance". In the most general sense, those attitudes are conditioned by the balance of good and bad associated with what is being received. Attitudes to risks imposed by major hazard plants will be conditioned by the balance of risks and benefits associated with them. This section looks at some of the factors affecting these attitudes. Subsection 6.2.1 considers how an appreciation of risk conditions its tolerance whereas subsection 6.2.2 covers benefits. Possible balance points between the two are described in general terms in subsection 6.2.3. A brief description of work on how risk is perceived in a more detailed way is given in subsection 6.2.4. Conclusions are drawn in subsection 6.2.5 as to the broad strategies available for evaluating whether risks are 'acceptable' or 'tolerable'.

6.2.1 Acceptance and appreciation of risks

Tolerance of risks is assured if those exposed to them are unaware of their existence. On the other hand, recognition that risks exist, but failure to appreciate their true characteristics, may be a major cause of intolerance.

A hazard may be 'accepted' unknowingly either because its existence has not been discovered (which we refer to as 'global ignorance') or because those subject to it have not been made aware of its existence or characteristics (which we refer to as 'local ignorance').

Global ignorance would imply that a hazard associated with some activity had yet to be discovered, or perhaps to be accepted in the scientific community. This might arise in any case where the risk associated with an activity was small, or difficult to link back to the cause. Many chronic health hazards fall or once fell into this category. Long term exposure to radiation was not recognised as harmful until the early pioneers succumbed to its effects; the link between smoking and lung cancer was only finally proven by careful epidemiological studies. Numerous other examples could be cited.

It can be anticipated that many more sources of hazards will be identified by future research. There are some cases where an unidentifiable potential hazard might provide grounds for restricting an activity (for example, some aspects of genetic engineering research). In the nuclear case, though, it seems unlikely that the range of hazards identified and discussed in Chapter 3 will expand in any qualitatively new directions in future; there has been extensive experience of human exposure to radiation to date, and its effects have been carefully monitored.

Local ignorance would imply that a hazard was recognised in some part of the community but that some of those exposed to risks from it were unaware of its existence. This concept might have sinister overtones; the idea that companies engaged in activities presenting hazards to the nearby population will maintain a state of ignorance of those hazards for profit motives is particularly repellent. A major purpose of UK and other health and safety legislation is to prevent this from happening.

Generally, though, local ignorance stems from completely innocent causes; the channels of communication used to inform people about potential hazards may be ineffective, or people may simply chose not to listen to genuine efforts to inform them. The possibility and effects (serious fire) of thermostat failures in domestic electrical appliances would be a typical example of a hazard of which some people are aware and take care to avoid while others are totally ignorant of its existence.

Although there is now almost universal recognition that there are hazards associated with radioactivity, it is very clear that the nature of those hazards is not well understood by most people. An important factor in the nuclear risk debate is that many opponents of nuclear power see the nuclear industry as trying to secure acceptance by concealing from the public the true nature of nuclear hazards. The combination of feeling 'kept in the dark' and poor understanding of radiation hazards provide ideal breeding conditions for intolerance of nuclear risks.

In summary, true ignorance of the existence of a hazard may, on a superficial level, appear to promote 'acceptance.' In the long run, though, a proper appreciation of hazards is an essential prerequisite for genuine tolerance.

6.2.2 Acceptance and appreciation of benefits

It has been suggested that a large part of the nuclear industry's failure to secure public acceptance derives from the lack of appreciation by the vast majority of people of the benefits arising from nuclear power. Associated with this is a lack of control; the judgement as to benefit has been taken remotely from those exposed to the risks. In these circumstances, any discussion of risks enhances the sense of their imposition on people without consent or compensation.

A particular difficulty is that many of the benefits of nuclear power - improved air and environmental quality, conservation of fossil fuels, diversity of electricity supply - cannot be measured on any widely accepted scale. Some of these strategic benefits are, indeed, matters of political opinion rather than hard fact. Others, such as the economic benefits, are open to quite widespread dispute, or depend on future values of parameters which cannot be reliably predicted.

This lack of a scale for measuring benefits, or even of a universally accepted framework in which to place those benefits, is a major factor hampering any attempts to secure acceptance through the evaluation of risks in relation to benefits in the nuclear context.

6.2.3 Risks and benefits - possible balance points

'True' acceptance of risk involves balancing risks and benefits of activities against each other. Consciously or otherwise, such balances are an everyday part of life. It is instructive to examine cases where people choose to accept risks, and the different levels of concern over risk which can arise. Three particular levels of concern are considered here: unconcern (distinct from ignorance of a hazard), concern or reluctance to accept risk, and willing acceptance.

Unconcern may arise from causes related to ignorance of hazards, though a distinction is made between ignorance of the existence of a hazard and awareness of its existence but lack of concern about the associated risks. Thus lack of experience of a hazard, either at first hand or via friends, family, or media, would be one explanation for a lack of concern. For many people, nuclear hazards were of little concern prior to the accidents at TMI and Chernobyl for this reason. Public concern in Britain over the risk of damage to life and property from high winds was minimal before the gales of October 1987. Similarly (by analogy with the 'ignorance' situation discussed above), concern may be dulled by the lack of direct and immediate links between cause and effect. This is one reason why chronic hazards, such as that of lung cancer associated with smoking, or of disease associated with occupational exposure to hazardous chemicals, cause many people relatively little concern.

Quite distinct from unconcern paralleling unawareness or ignorance of risks, though, is lack of concern arising from a judgement that a risk is in some sense small. The term 'de minimis' is often used to denote a level of risk below the smallest level of concern. Many attempts have been made to stipulate a basis on which risks may be described as 'de minimis'; this is a topic to which we will return in section 6.4. Finally people may fail to be concerned about a risk because of an attitude that it could not happen to them.

Concern or **reluctance** in accepting a risk may arise when there is either a greater awareness of the existence of a hazard whose effects are difficult to assess or consciousness of a higher level of risk from a hazard which can be assessed more confidently. Acceptance would then typically be conditional on some substantial compensatory benefit - the benefits of employment for someone accepting a job in a hazardous industry, or of danger money for merchant seamen in the Gulf, or of a package of community benefits for a neighbourhood accepting a major hazard plant. A wide spectrum of degrees of concern may be involved, and these situations are typically characterised by continuing negotiation with a view to improving the risk/benefit balance for the people exposed to risk.

Wholehearted acceptance of risk, for example by insurance companies, is characterised by a high degree of confidence that the benefits will outweigh the risks. This degree of confidence may be achieved in one of two ways. The first is by accepting only risks which can be quite precisely estimated. The second is by accepting risks which are more uncertain, but by demanding a relatively high benefit to cover the eventuality that risk might have been under-estimated.

Two major points emerge from the consideration of risks accepted voluntarily. Firstly, there are risks which are regarded as 'de minimis' - too small to be of concern. Secondly, in balancing risks against benefits, it is not only the estimated magnitude of the risk or benefit which matters, but the degree of confidence which can be placed in that estimate. Even where there is a well-defined scale for measuring risks or benefits, the value placed on that degree of confidence may significantly affect the balancing process.

6.2.4 Risk perception

The previous discussion has been very general. In practice a crucial feature bearing on whether risks will be tolerated is the detailed way in which they are perceived by various sections of society. A great deal of work has been carried out in this area, a close examination of which lies outside the scope of this report. This is not to say that we consider public perceptions to be unimportant in risk management, nor would we argue that this is not a topic which can be examined using scientific methods. But the question of how this can be done is sufficiently unresolved at present that in this report we simply provide a very brief overview.

Much of the early work in this area was done by Slovic, Lichtenstein and Fischhoff (1979, 1980, 1981). Work in the nuclear field has been carried out by Otway and Fischbein (1977) and Niehaus and Swaton (1981). A more recent review and critique of the methods employed is contained in the proceedings of a seminar organised by the authors of this report (Roberts, 1988).

General features which emerge from this work in relation to public risk perceptions are:

- Concentrated and obvious risks (motorway pile-ups, major industrial accidents) are regarded as worse than the risk from general road accidents or small scale industrial events even when equal numbers of deaths occur.
- Involuntary risks are regarded as worse than voluntary risks.
- Unfamiliar or new risks are regarded as worse than risks from familiar, natural or established hazards.
- Risks which are seen as inherently uncontrollable are regarded as worse than risks which can clearly be mastered.
- Risks evaluated by groups who are suspected of not being independent or competent are regarded as worse than risks evaluated by clearly impartial and competent groups.
- Immediate hazards are regarded as worse than deferred hazards.
- Risks from which there is no immediate personal benefit are regarded as worse than risks giving such a benefit. For example, off-site risks from nuclear power stations are regarded as worse than risks to on-site workers.
- Risks to named and clearly identified people are more important than risks in which no identified individual is under threat.
- Risks to an individual in a small group are less important than risks to an individual in a larger group.
- Risks over which the individual or her community have no direct control are regarded as worse than those for which there is some such control.

Although all these factors could in principle be evaluated in some quantitative way (for example by 'risk conversion factors' (Rowe, 1977; Litai, Lanning and

Rasmussen, 1983)), the absence of an accepted psychological model of how risk is perceived, as well as the objections listed in sub-section 8.2.2, suggest that this may not be useful at the present time.

6.2.5 Conclusions

People's attitudes to, and perceptions of risks are very complex, and do not of themselves provide guidance on the way in which the tolerability of risks should be evaluated. We can, though, identify three possible ways in which tolerance could in principle be established:

-risks will be endured if those exposed to them are unaware of their existence,
-risks might be shown to be below some threshold for concern, or
-risks might be shown to be outweighed by associated benefits.

In either of the latter two cases, confidence in the risk evaluation process would be just as important as any quantitative conclusions about risk values.

A risk management approach based on the first alternative can be rejected outright. Each remaining strategy brings its own problems. It is almost axiomatic that a hazard which is little understood yet of major concern cannot be assigned to the 'de minimis' set, which bodes ill for application of the second approach to nuclear power risks. Balancing of risks and benefits, though, cannot be done in any unique way; different groups will inevitably have different perceptions of the relative importance of different factors, and there is no obviously fair way to combine those perceptions.

The remainder of this chapter examines possible yardsticks against which risks might be evaluated, the ultimate target being to establish whether risks are either too small to be of concern, or are outweighed by associated benefits. We recognise that the development of a technical basis for risk evaluation is only a part (but an essential part) of a process for society to make decisions on tolerability of major hazard risks. In addition, this process needs to establish confidence in the risk assessment procedure and the standards used in risk evaluation, and assurance that public interests are being protected.

6.3 Strategies for the Evaluation of Risk Tolerability

In section 6.2 we identified two alternative conditions under which tolerance or acceptance of major hazard risks might be secured. In this section, we examine possible strategies for evaluating risks with a view to achieving either of these conditions. Subsection 6.3.1 looks at ways in which risks and benefits might be balanced against each other, while subsection 6.3.2 examines what would be involved in evaluating tolerability by looking at risks in isolation from benefits.

It is assumed in this 'strategic' section that, whatever the approach adopted to the evaluation of social risk tolerability, individual risks to people most exposed to major hazards will be controlled in some way and the discussion is focused exclusively on social risks.

6.3.1 Evaluation of risks against benefits

> ... there is no case for building Sizewell B unless a broad comparison shows that the benefits are expected to outweigh the risks with a reasonable degree of confidence (Layfield, 1987, para 36.95)

...these respective advantages and disadvantages belong to different categories and cannot be directly balanced one against the other (Inspector's report, Pheasant Wood inquiry)

These two quotations sum up the case for and against evaluation of social risks by comparison with benefits. We might easily draw up lists such as the following of benefits and risks associated with nuclear power:

Benefits	Risks
Conserve fossil fuels	Public : routine discharges
Economic electricity	major accidents
Clean air	Workers : occupational
Diversity of power supply	exposure
Avoid coal mining risks	accidents
Employment	Uranium miners, ...

While it is a useful exercise to clarify what exactly are the benefits and risks to be weighed against each other, the practical difficulties of doing this quantitatively are great. They are increased if it is desired to introduce some element to account for 'public perception' as described in subsection 6.2.4. The available methods and their difficulties, are described in Chapter 8.

However, in addition practical, though indirect, approaches for nuclear risk evaluation based on this philosophy could be developed. Two such approaches are considered here, the first based on the idea that nuclear power confers unique benefits, the second based on the idea that it confers benefits equivalent to those of other electricity generating technologies. These approaches would amount to political decisions to limit the range of options considered.

The 'Unique Benefits' Approach: In this approach, the benefits of, say, nuclear power are viewed as unique and essential. Since there are no other activities conferring these benefits, there is no reason to insist that nuclear risks should be smaller than risks of other activities. An extreme proponent of this viewpoint might argue that social risk management was thus unnecessary. More reasonably, social risk management might be based simply on the ALARA principle, that any cost-beneficial means of reducing risks should be implemented, but without any absolute yardstick against which to measure and evaluate social risks.

This is regarded as a reasonable strategy for social risk management. It would, though, be very difficult to secure agreement as to the uniqueness and desirability of the package of benefits associated with nuclear power.

The 'Equivalent Benefits' Approach: In this approach, nuclear would be regarded simply as one of a number of competing technologies for the necessary job of electricity generation. The principal benefit of each of these technologies - provision of secure, economic electricity - would be the same. If electricity generation is viewed as essential, risk management might reasonably be reduced to choosing from the lowest risk options among the competing technologies.

Again, this is regarded as an acceptable approach to social risk management. The principal problem is that the risks associated with different generating technologies are so different in nature. These differences are discussed further in subsection 6.5.1.

6.3.2 Evaluation of risks in isolation from benefits

In section 6.2 we discussed two useful concepts which could be used for evaluation of risks without direct comparison with benefits. The first is that of risks which are 'de minimis' - too small to be of concern. The problems with this approach lie firstly in defining a set of risks of no concern, and secondly in making a valid comparison between those risks and the risks of, say, major nuclear accidents. A particular problem is that levels of 'concern' do not correlate directly with numerical risk levels. Indeed, in the nuclear case, risks are 'of concern' almost by definition, so that procuring widespread acceptance of a risk management strategy along these lines may be difficult. The other useful concept is of a level of risk which is clearly too high to be tolerable, though this is not considered further in this chapter. The nature of both these levels and their implications for risk management are discussed further in Chapter 8.

The remainder of this chapter examines how such levels might be derived. There are several ways to do this including: comparisons with existing risk (though existing does not necessarily imply tolerable - see section 6.6); people's expressed preferences (though this might cause exaggeration of perception); or their revealed preferences. Because there is no unique way we shall be concerned not so much with determining intolerable or de minimis levels of risk as with defining reference levels which are useful to the risk evaluator without providing in themselves a decision rule.

6.4 Possible Reference Levels for Societal Risk Evaluation

This section examines how reference levels of risk for use in risk evaluation can be derived. Attention is focused on nuclear risk management, though the principles can be applied in any context. Subsection 6.4.1 provides a general discussion of possible levels of risk against which nuclear risks might be evaluated. Many attempts to compare nuclear and non-nuclear risks have been heavily criticised as failing to compare like with like, or avoiding important issues. Hence, subsection 6.4.2 discusses the general principles which apply to the comparison of risks arising from different sources.

6.4.1 Candidates for comparison

There is no simple, fundamental principle from which a small level of risk in an absolute sense can be determined. The only way in which to provide some perspective on the magnitude of nuclear social risks is to compare them with other risks. While the combination of hazards associated with large radioactivity releases into the environment is probably unique, many of those hazards are encountered, separately or in different combinations, in various other types of activity. We consider four possible candidates for providing levels of risk against which to evaluate nuclear social risks:

- -social risk levels associated with alternative means of electricity generation,
- -social risk levels related to standards for individual risk,
- -existing social risks associated with background radioactivity, and
- -other existing social risks.

The degree to which levels of risk comparable with those derived from the above sources can be termed 'small' is an important part of the discussion. The first three candidates for comparison contain an element of 'smallness' in that they represent risks which are already tolerated in some sense. Such comparisons are discussed in section 6.5. The fourth is different. In fact many of the existing risks with which we might make comparisons are very definitely above thresholds for concern. There is no obvious way to define a level which is not of concern by direct comparison with such risks. Some possibilities are discussed in section 6.6. However, no judgement is made in this chapter as to how small a level must be to be 'of little or no concern to ordinary people.'

6.4.2 Principles for risk comparisons

These principles represent an attempt to itemise ways to ensure that risk comparisons are as fair as possible. They cover both the scope of comparisons (how different hazards should be covered, and what risks should be considered on either side of a comparison) and the equivalence of risks associated with different activities (how comparisons can be made fairly). The principles are presented as a set of axioms, each followed by an explanation of its intent.

> **Principle 1**: Separate standards should be developed for different types of risk. For activities presenting a range of different types, the associated risks may then be evaluated separately.

Attempts to equivalence qualitatively different types of risk (for example, expressing injuries as equivalent to fractions of a death, or early deaths as equivalent to a number of late deaths) should be discouraged unless a clear and unequivocal basis for such equivalence can be demonstrated. Such equivalence may ultimately have to be factored into a decision process, but the technical part of the risk evaluation process should avoid value judgements of this sort.

For major nuclear accidents, it is therefore suggested that the types of personal and non-personal risks itemised in section 3.2 should be assessed separately. This means that there are six types of risk for which standards are sought in sections 6.5 and 6.6: personal risks (early death, delayed death from cancer, serious injury, permanent evacuation and serious birth defects) and non-personal risks (mainly financial).

We focus our attention primarily on the personal types of risk. Although we recognise that the non-personal risks may be an important part of the societal impact of major accidents, financial risks are considered to be of lower priority in the debate over nuclear risk tolerability.

> **Principle 2**: Major hazard risks should be compared with risks that are not voluntary.

We recognise that it may be difficult or impossible to draw a clear distinction between voluntary and involuntary components of risk. For example, risks associated with work, transport and diet are all nominally within our control, but are for practical purposes fixed for many people. Smoking is a voluntary activity which causes many lung cancer deaths, yet lung cancer can be caused by other factors than smoking, and we cannot attribute a certain fraction of lung cancer deaths to smoking. So, even where cause and effect are established, it may not be practicable to subtract a portion of existing risks corresponding to voluntary activities.

Some risks, though, such as those associated with leisure activities, or, in the extreme, suicide, are clearly voluntary and should if possible be excluded from any comparisons.

The separation of risks to workers and to the public is a useful area of application of this principle. Thus, for example, multiple fatality risks to workers in the chemical industry should not be used as a basis for comparison with nuclear multiple fatality risks to the general public.

A possible extension to this principle might be to exclude from comparisons specific groups of the population at very high risk. For example, accidents in the home are a major cause of death for the very elderly, but involve relatively modest levels of risk for the rest of the population. When comparing with an accident hazard affecting the whole population indiscriminately, it might be argued that such groups should be excluded from existing risk totals. While this is a potentially very important factor in the evaluation of risks to most-exposed individuals, it is less important for evaluating risks integrated over society.

> **Principle 3**: Qualitative differences between risks involving similar consequences arising from different sources should be identified and, so far as possible, allowed for in making comparisons.

There are often significant differences between nominally equivalent hazards arising from different sources. Factors often encountered when comparing nuclear and non-nuclear risks include age at time of death, the degree of uncertainty attached to risk estimates, the different levels of existing risk for different groups in society, and the different perception of single accidents involving many people as opposed to many accidents involving one or few people each. These factors may influence the choice of a measure of risk as a basis for comparison; some of the implications are discussed below.

Age at time of death If we define a consequence as 'incidence of premature death,' that consequence would be of greater concern if it affected primarily young people than if it affected solely the very elderly. The disease AIDS, although it kills far fewer people, causes greater loss of life expectancy (LLE) than many more familiar diseases. Consideration of LLE (see Chapter 3) provides a partial solution to this problem. For example, this is a useful measure for comparing the integrated societal risk of early death from nuclear accidents with the existing fatal accident risk (which affects the elderly more by comparison) as discussed in subsection 6.6.1. By contrast LLE is not a good way to equivalence early fatalities (as has often been proposed) because of qualitative differences between the two consequences (the difference between a stochastic and a non-stochastic risk, the non-attributability of cancer death to irradiation, the different degree of uncertainty associated with the two types of risk; see subsection 6.6.2).

Based on a recent model of the health effects of ionising radiation (USNRC, 1985), indicative calculations have been performed of the average loss of life expectancy associated with early and late fatalities for a uniformly exposed population with age and sex cohorts corresponding to those of Great Britain. Equivalent calculations of 'loss of life expectancy per death' have been performed for existing risks (actuarial risks of accident and cancer fatalities), and predicted risks of cancers associated with natural background radiation. The results are shown in Table 6.1. Figures such as these provide a valuable adjunct to numbers of fatalities, particularly for risks integrated over the whole population.

Groups at risk The AIDS example illustrates another relevant factor mentioned under Principle 2; that is, the ability (for some effects) to distinguish a subset of the population which is at high risk, while others are not at risk at all. Many small minority groups could be identified with very high levels of risk from specific causes (for example, North Sea divers, drug abusers, participants in dangerous sports), but many of these are voluntary. Moreover, though leading to high levels of individual risk for those involved, such activities generally contribute little to the overall incidence of death and disease in society.

Of more concern are the differences in existing risk between much broader groups in society. Differences between the sexes are an obvious case in point. Women generally experience lower risks of accidents than do men. Some fatality risks (for example, cancers of breast or prostate) are not shared by the sexes, while others (for example, lung cancers, many occupational and transport risks) are encountered to a far greater extent in men than in women because of differences in habits, rather than physiology. Although the differences between aggregated cancer and accident risks are not great for men and women (no more than factors of two or three), many of these differences are consistent and significant. Comparative risks are thus presented for the sexes separately where practicable.

Multiple fatality risks The concept of 'high consequence aversion' (see Chapter 2) has been much discussed, and is incorporated into several proposed safety targets for nuclear and non-nuclear industries. All of the arguments for the existence of this phenomenon are based on aversion to large numbers of prompt fatalities attributable to single events. It is debatable whether it is the single cause or the collocation of those affected (impact on a single community) which is primarily responsible for the disproportionate aversion to multiple fatality accidents. Weather systems quite frequently account for multiple fatalities (for example, gales in January 1976 killed at least 24 people (Buller, 1977); the gales of October 1987 killed 19 (Sunday Times, 1987)), yet it is arguably the widespread damage to property and the environment, rather than the multiple fatalities, which focuses so much attention on such events.

Whether commonality of cause or of location is responsible for 'high consequence aversion,' it seems reasonable that any such concept associated with other existing risks should apply equally to nuclear early fatality risks, and possibly to other personal risks in the immediate aftermath of an accident. Other than by comparison with existing risks of multiple fatality accidents, though, we can deduce no satisfactory principle for establishing a numerical representation of 'high consequence aversion' (for example, a slope on a logarithmic CCDF) for incorporation into standards for risk evaluation.

We do not, however, consider there to be any grounds on which the concept of 'high consequence aversion' can be associated with predicted multiple late cancer fatalities arising from nuclear accidents (or indeed any other stochastic risk). Any such fatalities would occur over a long period of time, would not be specifically attributable to a particular event or cause, and, if at all detectable, would be manifest as a small increase in the large background incidence of cancer spread out over a large group of people. In terms of numbers of deaths and distribution of individual risk, there would be absolutely no difference between ten accidents each involving an average 0.1% chance of death for a group of people and one accident involving an average 1% chance of death for the group (or more accurately an average 0.9955% chance of death - this is the chance of not surviving ten events each carrying a 0.1% chance of death). In principle the larger accident might be more detectable, but in practice this is unlikely to be the case for nuclear accidents.

It follows that comparisons between nuclear and existing societal cancer risks might therefore quite justifiably be made simply in terms of the expected numbers of deaths per year (but see discussion on uncertainty below). This is in contrast to the case for early death where the CCDF representation is necessary for any evaluation of 'high consequence aversion'. When societal risk of delayed cancer is factored into risk management policy, though, it is possible that it could nevertheless be desirable to formulate risk targets for delayed fatalities in terms of a CCDF. This might be done for several reasons: to recognise public aversion to large accidents given that total figures are available from PSAs and post-accident media reports; to incorporate as good practice this idea that all the risk should not be attributable to a single accident or group of accidents; or to control large accidents generally. These more arbitrary aspects of risk management are discussed further in Chapter 9.

Uncertainty Another very important attribute which may differ widely between an assessed and an existing risk is the degree of confidence attached to the estimate of each. Historical risks, especially for more common events, may in some cases be established with a very high degree of confidence; in others, absence or poor quality of data may make actuarially derived risks highly uncertain. Estimates of existing risks associated with events which have not yet occurred (for example, very severe weather, meteorite impacts, or accidents with existing technology) would be of particular interest for comparison with major accident risks but are largely unavailable or highly uncertain. The uncertainties in estimating risks for major hazard plants are described in Chapter 5.

A good example of the impact of uncertainty on the quality of the risk comparison is the prediction of large numbers of delayed cancer fatalities associated with exposure of large populations to small levels of radiation. A dose of 1 Sv to each of 100 people is qualitatively different from the same collective dose spread uniformly over 1,000,000 people in that there is direct evidence to support a link to delayed cancers in the former case, but not in the latter.

The qualitative differences between assessed risks and existing risk levels are as important to a decision maker as the quantitative comparisons themselves. It is therefore recommended that, wherever possible, risks should be presented and compared in such a way as to enhance the evaluator's appreciation of differences in 'texture' of different types of risk. In the case of the specific example above, presentation of collective dose in bands of individual dose or dose rate would provide the risk evaluator with a more complete picture of the nuclear risk being compared. In any case, as full as possible a description of any qualitative differences between risks should be provided with any comparison.

6.5 Evaluation Against Tolerated Risks

This section discusses numerical levels of risk which are tolerated in some sense. These are the first three of the candidates defined in subsection 6.4.1:
- -risks of competing technologies (subsection 6.5.1),
- -risks compatible with individual risk principles (6.5.2), and
- -risks associated with natural background radiation (6.5.3).

In each case we apply the principles for comparison laid out in subsection 6.4.2.

In comparing risks integrated over a population, the size of the populations considered is clearly important. Wherever possible, social risk levels are presented as integral totals over the population of Great Britain (England, Wales and Scotland), which is where all current UK nuclear plants are located. In some cases, information is available for the UK as a whole (including Northern Ireland and the Channel Islands) or for parts of the population. In such cases, totals for Great Britain have been estimated pro rata according to population.

Because the risk levels presented apply to the nation as a whole, they provide a basis for comparison with nuclear risks arising from the whole UK programme, and not for comparison with risks arising from a single plant or site.

6.5.1 Risks of competing technologies

The 'equivalent benefits' approach to tolerating risk described in subsection 6.3.1 bypasses the need to evaluate the benefits of nuclear power. One reasonable standard against which to evaluate nuclear power risks is that those risks should not exceed those of other means of generating electricity. The principal disadvantage is that the hazards presented by nuclear and non-nuclear power are not the same, as we shall see.

Many studies have been made attempting to compare risks of nuclear and alternative generating technologies. Coal and nuclear are the principal competing technologies in the UK, and the discussion here is limited to a comparison of the two. This is sufficient to provide an appreciation of the advantages and limitations of this approach to the evaluation of nuclear power social risks.

Rose and Billington (1987) have recently reviewed over fifty studies of risks associated with electricity generation from coal. A summary of the findings is contained in Table 6.2. Several features should be noted:

- All of the accident risks presented are of early, rather than delayed death.

- The fatality risks associated with disease are of a more long-term nature, but are due to prolonged exposure to normal operational discharges, not to accidents. The fatality risks from chronic exposure to the oxides of sulphur and nitrogen at very low levels are highly uncertain.

- The public fatal accident risks are primarily associated with transport and would therefore be borne primarily by road users, a population weighted towards younger males. These risks are arguably part of the spectrum of transport risks accepted voluntarily by road users (see subsection 6.6.1).

- The public fatal disease risks would be spread over the entire population, male and female, but fatalities would be concentrated in those most susceptible to bronchial complaints, which would weight fatalities heavily towards older age groups.

This table, like most other comparative studies of coal and nuclear risks, considers only fatality risks. Comparisons with such risks should take into account the different populations affected by coal and nuclear risks and the different nature of, and uncertainties associated with those risks. Before summarising coal social risk levels for the Great Britain population, it is worth considering some of the other aspects of social risks associated with coal which are not covered in Table 6.2.

Multiple Fatality Risks The possibility of multiple fatality accidents affecting the public can be identified for at least the last three of the stages in Table 6.2. Examples include accidents involving chlorine transport and storage, other road accidents, generation of turbine missiles, and landslip of spoil heaps, as occurred in the Aberfan disaster. Other types of multiple fatality accident have occurred elsewhere in the world; in the early 1970s around 250,000 people were killed in sudden, widespread subsidence in Manchuria.

Some of these hazards, having been recognised over the last decade or two, have now been more or less eliminated and in current UK conditions, it is very difficult to conceive of accidents in the coal industry which could cause large numbers of deaths. In this respect coal and nuclear hazards are not directly comparable.

Non-Fatality Risks Although these are not exactly of the type associated with large nuclear accidents, there are many adverse social impacts of the coal industry, which we consider here under the various categories of personal and non-personal risk with which we are concerned.

Injuries would presumably be produced in parallel to fatalities in much the same way as for other transport risks, and would likewise be associated with multiple fatality accidents. There is no available data on incidence of injuries to the public associated with electricity generation from coal.

Temporary and Permanent Evacuation are among the most commonly encountered and serious social impacts of the coal industry, in association with subsidence in mining areas. A discussion of the impacts of subsidence is given in a report by the Commission on Energy and the Environment report (CEE, 1981). Subsidence affects buildings, pipes, sewers, cables and land use generally. It can involve both temporary 'evacuation' from dwellings during reconstruction, and essentially permanent effects on land use.

Serious birth defects There is no evidence to link generation of electricity from coal with birth defects.

Non-Personal Impacts Subsidence damage does not only involve personal impacts such as forced relocation, but also non-personal impacts, including the financial compensation mentioned above as well as factors such as diversion of transport routes, speed restrictions on railway lines, and permanent flooding or waterlogging of land, with all its implications for land use and ecosystems. As with many of the fatality risks, though, these non-personal risks are associated more with controlled, normal operations rather than with accidents.

Scaling from Table 6.2 based on 20 GW-years per year generation produces the social risk estimates shown in Table 6.3 for generating electricity from coal for Great Britain. The table shows only the public risk and not the occupational risk (fatal accident, 28 per year; fatal disease, 6 per year) since the latter is considered voluntary. We suggest that the coal social risks be scaled down in proportion to the quantity of electricity generated before making a direct comparison with nuclear risks; the scaling factor would be about four at present.

The risks of electricity generation from coal considered above do not include several potentially very serious ones which have not so far been estimated with any accuracy. Two examples are the risks from earthquakes in areas liable to subsidence and the so-called greenhouse effect. This last arises from increased levels of carbon dioxide in the atmosphere and the consequent increased temperatures leading, perhaps, to hundreds of millions of casualties. However, such risks are tolerated essentially because of unawareness, and they do not provide useful reference levels or risk targets.

Attractive though the idea of evaluating nuclear risks against competing energy technologies is, the lack of direct comparability of the hazards presented by nuclear and other means of electricity generation, and the qualitative differences between those risks which do appear comparable, are considered to cause great difficulties for comprehensive, quantitative evaluation of nuclear social risk acceptability by this means.

6.5.2 Social risk standards from individual risk standards

A level of individual risk of death of 10^{-6} per year is widely regarded as being below the level at which most ordinary people are concerned about hazards, and has indeed been adopted as a standard for protection against accident risks of most-exposed members of the public by the UKAEA (1987). It might at first sight be thought that control of individual risk at such levels would also control social risk in a suitable way. However, because these targets refer to maximum levels of individual risk, it can be anticipated that the average individual risk will be very much smaller. Furthermore, it should be remembered that society as a whole is at risk from a large number of activities, whereas individuals (or at least those with risks approaching the target) will probably have their risk dominated by one major hazard site. It follows that social risk must be controlled using different principles from those that control individual risk as described in Chapter 2.

The concept of average individual risk is also one which has been proposed for use in societal risk management (see the US Safety Goals in Chapter 7). However, when it is realised that adding more individuals to those locations within the region where the risk is lower than average actually reduces the average individual risk, while increasing the social risk, it is clear that such a quantity does not well control social risk.

6.5.3 Risks from background radiation

Background natural radiation from terrestrial and cosmic sources, and to a lesser extent radiation associated with human activities (principally medical uses of radiation) gives rise to annual effective doses to everybody in the UK of the order of 1 to 100 mSv per year. Regional variations of the order of 1 mSv per year and more appear to be of no concern to the public, and it might reasonably be argued on this basis that social risks corresponding to these levels of exposure of the population at large are below some level of concern. Thus, existing radiation is a very attractive yardstick against which to evaluate stochastic nuclear risks (delayed cancer fatalities, genetic effects) where it is perhaps the most directly comparable existing risk with nuclear power risks.

The above levels are, however, generally too small to give rise to effects which would occur only above a threshold dose (early fatality, evacuation, interdiction of land use and agricultural produce). Countermeasures to mitigate natural radiation exposure are under consideration, though, for a small number of homes with high radon levels in both the UK and the USA, and it is interesting to note that natural radioactivity in several imported foodstuffs exceeds some of the more restrictive limits adopted by some European countries in the wake of the Chernobyl accident.

We therefore consider qualitative differences between background radiation and nuclear accident risks only in the context of stochastic health effects. Several points are relevant:

- The small doses involved and very low rate at which such doses are received has both advantages and disadvantages in comparing with delayed fatality risks from accidents. A substantial proportion of the committed collective dose from large nuclear accidents (and all of it for many smaller ones), will, like background radiation, be associated with individual dose levels of an order below the 0.1-1 Sv range where there is direct evidence linking excess cancer deaths to radiation.

  Some would argue that this makes background radiation an ideal candidate for comparison with nuclear accident delayed fatality risks. It should be remembered, though, that in large accidents significant collective dose would be contributed by individual doses at considerably higher levels.

- Whereas background radiation is accumulated over an entire lifetime, radiation dose following a nuclear accident would be accumulated over a shorter period (typically about half in the first year, with an amount decaying by a factor of two in each subsequent year). This means that, for a given dose commitment, predicted fatalities following an accident would involve proportionately more fatalities among younger people and fewer among the elderly. This is reflected in the loss of life expectancy (LLE) figures summarised in Table 6.1 which shows the effect not to be large.

- Certain components of background are at least to some extent voluntary. Hughes and Roberts (1984) report the proportion of UK collective background radiation exposure deriving from various sources. Some artificial sources, such as medical and occupational exposure, might well be classed as voluntary, while all of the natural sources are to a substantial extent avoidable (by living at sea level in well-ventilated wooden houses in areas of low terrestrial activity!) Being more realistic, medical, occupational and, perhaps, radon exposures are those with some claim to being avoidable; these would imply a factor of far less than two adjustment.

The total collective dose is 120,000 manSv per year. This dose may be converted to a corresponding number of late fatalities and genetic defects using the ICRP factors reported in Chapter 3 and the results are shown in Table 6.3. (Note: the more recent radiation health effects model (USNRC, 1985) used to generate Table 6.1 predicts a 30% higher cancer fatality rate.)

In summary, background radiation exposure seems unlikely to be of use in formulating risk targets, but may have application as a risk comparator.

6.6 Evaluation Against Other Existing Risks

Many aspects of existing risk were reported in Chapter 4. The use of existing risks generally as a reference scale for major accident risks is fraught with difficulties in ensuring that like is being compared with like, that fair comparisons are being made, and in deciding what constitutes a 'small,' or a tolerable level. In general we agree with the Warner report quotation cited by the Friends of the Earth at the Sizewell Inquiry in support of their claim that it was impossible to infer the tolerability of new risks from the tolerance of existing risks:

> The existence, or indeed the informed acceptance of any set of pre-existing risks, provides absolutely no justification by itself for imposing any other hazard even when the risk is reduced as far as it reasonably can be..... In general, an additional activity can be justified only by reference to its associated benefits and the cost of associated risk reductions.

Nonetheless, this approach provides unique opportunities for placing nuclear accident risks in an overall perspective, and we consider it an important source of possible reference levels against which accident risks might be evaluated.

The six categories of social risk identified in Chapter 3 are considered separately: early fatalities (subsection 6.6.1), delayed cancer fatalities (6.6.2), serious injuries (6.6.3), evacuation (6.6.4), birth defects (6.6.5) and non-personal risks (6.6.6).

In each subsection, existing risks considered to provide a possible comparison with nuclear accident risks are described. The qualitative features of those risks which hamper comparison with nuclear risks are discussed.

Approaches to defining 'small' levels of risk when the comparators are clearly of concern include:

- specification of arbitrary fractions of existing risks (for example: 10-100% as 'significant', 1-10% as 'small', 0.1-1% as 'very small'),
- levels comparable with regional or temporal variations in existing risks, and
- consequence levels too small to be detectable (for stochastic health effects only).

However, we recognise that levels derived using these descriptions can be indicative only.

6.6.1 Early fatality risks

Two types of existing risk are considered: those for events which have occurred and on which historical data is thus available (termed 'actuarial' risks), and those for events which have not occurred but for which predictions have been made (termed 'predicted' risks). This is the only risk category for which predicted risk estimates are available for sources other than major hazard plant. We begin by considering the actuarial risk of accidental death.

Figure 4.1 shows the risk of death in Great Britain from a variety of causes classified as "Injury and Poisoning", but omitting suicide (OPCS, 1985). It is this category which provides the best comparison with nuclear early death risk. CCDFs giving the frequency of multiple fatality accidents in the UK are presented in Figures 4.2 (various causes), 4.4 (chemical industry) and 4.5 (aircraft accidents).

We discuss the qualities associated with these various hazards before considering what might be interpreted as a small level of risk. The qualities are: comparability, age at death, voluntary/involuntary distinctions, groups at risk, multiple fatalities and variability.

Comparability Although the risks quoted above appear to be the most suitable for comparison with the risk of early death from nuclear accidents it is clear that the nature of death over days or weeks from a large dose of radiation is qualitatively different from death in an aircraft accident, for example.

Age at death As can be seen from Figure 4.1, there is a very wide variation of accident fatality risks of most types with age. In general, transport risks are greater for younger people, while accidents in the home are most important for the very young and, particularly, for the elderly. Overall, accident fatality risks increase with age, with the over-65s accounting for about 38% of deaths yet only about 15% of the population. This explains the two peaks in the figure.

In contrast, early fatalities following a major nuclear accident would be expected to strike fairly indiscriminately across the population. (In fact, there is some expectation that children would be more sensitive to radiation than adults, but there is insufficient data at present to quantify this (USNRC, 1985)).

Thus the LLE associated with 'nuclear early death' would be greater than that associated with existing risks in society: Table 6.1 presents estimates of 15 and 39 years (population average) life expectancy lost per death for existing accident risks and nuclear early death risks respectively.

Voluntary/Involuntary distinctions Suicides have been omitted from Figure 4.1 because they are clearly voluntary in one sense (though most result from involuntary depressive illnesses), and most of the other 'deliberate' categories (including the large numbers of "not sure whether self-inflicted") are doubtful.

Furthermore, there is a large voluntary component to transport risks, evidenced by the high risk to young males. Removing all these categories from the total reduces the average risk by a factor of 2.6.

Groups at risk The main point is the higher risk to men than to women at all ages - by a factor of 4 at age 20-24. And, as an extreme case, a category which also excluded transport risks would show that young girls' risk at around 10^{-5} per year is an order of magnitude lower than the pre-retirement adult population mean.

Multiple fatality risks The data are only available for accidents up to around 100 fatalities. Furthermore, there is a very large voluntary component. Conversely, there is a clear association of serious accidents with modern technology (in the UK) which strengthens the argument for comparison with nuclear risks. It should be noted that road accidents are the dominant contributor for accidents with less than 10 fatalities. The data show $F(1)=10^4$ per year, $F(10)=1\text{-}10$ per year and $F(100)=0.1\text{-}1$ per year (where $F(N)$ is the frequency of accidents killing N or more), but lack of data and comparability mean there is no real basis to define a reference CCDF. This situation might be alleviated by using world wide data, but in this case the basis for comparability is doubtful because of cultural differences.

Variability The regional variability of the accident risks is available in Scottish data (RGS, 1985). The variations in individual risk are of the same order as the risk itself. This supports the description of 10% of existing risks as 'significant' as proposed above.

In summary, the actuarial social risk of accidental death is about 14,500 deaths per year, excluding suicides only. It is not possible to construct a CCDF which provides a useful reference level for risk evaluation. This figure could be modified by a factor of 2.5 to amount for LLE, and up to 2 to remove the voluntary component.

Turning now to predicted or estimated existing risks, there are several good candidates for comparison at high consequence levels. One risk which is frequently cited is that from meteorite impacts. Another is that from extreme weather conditions; the data for the UK is rather poor, but it can be anticipated that because structures are typically designed to withstand one in 50 year events, damage will escalate rapidly above this return period. This appears to be a promising area for further work, though present data is insufficient to define useful reference levels.

6.6.2 Delayed cancer fatalities

Figure 4.1 also shows the risk of death from cancer in Great Britain. This provides the best comparison with the nuclear cancer risk from accidents. As stated previously we are interested only in total numbers per year: 'high consequence aversion' is meaningless for stochastic risks.

Comparability These deaths are not from accidents, but otherwise the nature of the consequences is essentially the same.

Age at death The proportion of cancer deaths among children is very small; this would also be the case with post-accident nuclear delayed deaths on current models. However, indicative calculations do show this proportion to be larger (up to a factor 10 for some ages and cancers) for radiation induced cancer compared with the cancer statistics. In terms of LLE there is very little difference for the population average, as shown in Table 6.1. Indeed there is not much difference for

LLE among children, since death within a few years is a small fraction of the total number of deaths induced by irradiation at an early age according to the model used (USNRC, 1985).

Voluntary/involuntary The major causes of death from cancer are lung cancer (to a large extent caused by smoking) and cancer of the gastro-intestinal tract (for which there are large national differences caused by lifestyle and diet). There are also substantial avoidable components among the other cancers. Thus it might be conservatively estimated that as few as 10% of cancers are unavoidable.

Groups at risk There are obvious differences between the sexes arising from physiological causes (breast, cervix, prostate and so on) and also differences in habits (for example, smoking). Furthermore, there are differences with regard to age group; for example, only leukaemia and bone cancer are significant for children. However, these patterns for radiation induced cancer will in many cases be similar.

Variability Again the Scottish data show regional variability comparable to the size of the risk.

Detectability Because of the regional variability, and the random temporal fluctuations in the number of deaths, any additional cause of cancer would have to cause large numbers of deaths to be detectable. However, it is important to consider the most sensitive combination of disease and age group; this is where the higher incidence of nuclear compared with existing cancer for children might be particularly relevant. A undetectable risk is not, of course, a tolerable one.

In summary, there are about 150,000 deaths per year from cancer, of which not less (and probably much more) than 10% are unavoidable. Children are particularly susceptible to nuclear accident cancer compared with the existing cancer risk. Finally, because the variations in the numbers of cancer deaths are very large, hundreds to tens of thousands of deaths would be undetectable if spread over many years.

6.6.3 Injuries

Chapter 4 presents some information on occupational injuries, but there appear to be no available data for the public, although the information is in principle obtainable from confidential medical records.

In these circumstances the best approach appears to be to control injury using early deaths, though radiation induced injuries would be much more like chronic diseases than injuries sustained in accidents. For example dose thresholds for early death could be reduced by a factor (around 4, say) so that injury counted as death. This would in any case lie within the uncertainty of releases and consequent exposure.

6.6.4 Evacuation

The CCDF information for evacuation presented in Chapter 4 is much better for this consequence. However, in the UK evacuation is invariably temporary. There is also a marked contrast between nature of nuclear and non-nuclear land interdiction in the UK cases (for example, subsidence causes flooding and sterilisation of land, but does not preclude use as nature reserves, say).

Thus, although there are quite high frequencies for moving large numbers of people, it is not possible to define a useful reference level.

6.6.5 Birth Defects

Some 2% of all babies born in England and Wales suffer some form of congenital malformation (OPCS, 1987), of which some 20% prove fatal, half within the first year. The types of malformation range from relatively minor problems to conditions resulting in large loss of life expectancy (for example, anencephalus and spina bifida, both of which are showing marked reductions in time) and major alterations in quality of life (for example, Down's Syndrome).

No effects of radiation on human birth defects have been observed to date. Effects have been observed, though, in populations of mice and lower animals. Models of damage and repair have been inferred from these laboratory tests, and applied to humans to estimate the incidence of various broad classes of defect. Such models have been used to predict the expected incidence of birth defects among, for example, children of the Japanese atomic bomb survivors. With no observed excess, the best that can be said is that the models are not inconsistent with what data is available.

With these provisos in mind, we can compare the predicted effects of radiation in terms of birth defects with the existing incidence of birth defects using the Harvard model (USNRC, 1985); the figures quoted are regarded as indicative and do not imply endorsement of this particular health effects model. It should be noted that, like delayed cancer, induction of birth defects by radiation is considered to be a stochastic process.

The model predictions are that for a population of a million people uniformly irradiated with 0.01 Gy, 30 excess defects would occur in the first generation, and 185 excess defects would occur in total over all future time. Comparisons with the existing rate are complicated by the various classes of defect of varying severity. The same models, for the same exposure, predict about 180 excess cancer fatalities. The ratio of radiation induced to existing effects is about 1:1000 for delayed cancer fatalities, and 1:300-1700 for birth defects (depending on what classes are included).

The extreme difficulty in predicting radiation-induced genetic effects militates against attempting to make direct comparisons between nuclear and existing risks in this context. However, the broad similarity between birth defects and delayed cancer, as regards both the stochastic nature of the hazard and the excess effects produced by radiation in relation to the normal incidence of effects, suggest that any risk management strategy which controls delayed cancer fatality risks will provide a broadly comparable degree of protection against risks of birth defects.

6.6.6 Non-personal risks

It is possible to find data in this area; two topical examples are the frequency of large stock market losses and the costs of extreme weather. However, the financial consequences of accidents can be evaluated using available techniques and there is no requirement for reference levels of risk for such consequences. These methods do not readily deal with macroeconomic effects such as the impact of a nuclear accident on the nuclear programme, say. These non-personal costs were described in subsection 3.2.2, where we pointed out that such aspects lay outside the scope of this report.

6.7 Conclusions and Recommendations

(1) Tolerance of risks can in principle be established in two ways: they might be shown to be below some threshold for concern; or they might be shown to be outweighed by the associated benefits. In both cases it is possible to look for useful reference levels of risk.

(2) Our principles for risk comparisons state that: different types of risk should be evaluated separately; voluntary components should be excluded; and quantitative differences between the risks to be compared should be identified and, if possible, allowed for.

(3) Although 'high consequence aversion' could in principle be applied to early deaths we have found no satisfactory theoretical way to establish a numerical representation such as a CCDF shape. Cancer risks should, however, be treated simply in terms of the expected total number of deaths (subject to the associated uncertainty).

(4) Lack of direct comparability of the hazards, and qualitative differences between the types of risk which do appear comparable, cause great difficulties for comprehensive quantitative evaluation of nuclear social risk against that of competing technologies for electricity production.

(5) Reference levels of the social risk of cancer from nuclear accidents may be derived either from the existing cancer risk, or by comparison with background radiation. When compared with the existing cancer risk, the number of effects would have to be very high to be detectable.

(6) Comparison of the risk of early death from nuclear accidents with the existing accident risk is possible with a number of provisos. However, it is not possible to construct from the data a CCDF which provides a useful reference level for risk evaluation. The predicted risk from natural events may be helpful here.

(7) Injury and birth defect risks from nuclear accidents may be managed as part of a strategy for managing the risk of early and delayed death respectively.

(8) On the basis of historical data, we can find no useful reference levels for evacuation and non-personal risks.

6.8 References

Buller, P S J, 1977, **Gale Damage to Buildings and Structures in the United Kingdom, 2 January 1976**. Note CP 42/77, Building Research Establishment, Garston.

CEE, 1981, **Coal and the Environment**. Report of the Commission on Energy and the Environment, HMSO, London.

Hughes, J S, and Roberts, G C, 1984, **The Radiation Exposure of the UK Population - 1984 Review**. NRPB R173.

Layfield, F, 1987, **Sizewell 'B' Public Inquiry Report**. HMSO, London.

Litai, D, Lanning, D D and Rasmussen, N C, 1983, **The Public Perception of Risk**. In 'The Analysis of Actual versus Perceived Risks'.

Niehaus, F, and Swaton, E, 1981, **Public Risk Perception of Nuclear Power**. Colloquium on the Risks of Different Energy Systems, SFEN, Saint Etienne Cedex.

Otway, H J and Fischbein, M, 1977, **Public Attitudes and Decision Making.** International Institute for Applied Systems Analysis, Laxenburg, Austria.

OPCS, 1985, **Mortality Statistics, Cause: Review of the Registrar General on Deaths by Cause, Sex and Age in England and Wales, 1984.** Office of Population Census and Surveys Series DH2 no.11, HMSO, London.

OPCS, 1987, **Congenital Malformations and Monitoring System 1986.** OPCS Monitor, reference MB3 87/1, Office of Population Censuses and Surveys, London.

RGS, 1985, **Annual Report of the Registrar General for Scotland, 1984.** Government Statistical Service, Edinburgh, HMSO.

Roberts, L E J, (Ed) 1988, **Risk Percention and Safety Targets for Major Accidents.** Report of a seminar held at the University of East Anglia, 16 October 1987. Research Report No 4, Environmental Risk Assessment Unit, University of East Anglia, Norwich.

Rose, K S B, and Billington, D E, 1987, **Studies on the Risk of Electricity Generation: a Comparative Assessment. Class A: Coal.** Report AERE R 12057, UKAEA Harwell.

Rowe, W D, 1977, **An Anatomy of Risk.**

Slovic, P, Fischoff, B and Lichtenstein, S, 1980, **Perceived Risk.** In "Societal Risk Assessment: How Safe is Safe Enough" (R C Schwing and W A Albers, eds). Plenum, New York.

Slovic, P, Fischoff, B and Lichtenstein, S, 1981, **Perceived Risk: Psychological Factors and Social Implications.** Proc Roy Soc, A376, pp17-34.

Slovic, P, Lichtenstein, S and Fischoff, B, 1979, **Images of Disaster: Perception and Acceptance of Risks from Nuclear Power.** In "Energy Risk Management" (T D Goodman and W D Rowe, eds). Academic Press, London.

Sunday Times, 1987. Colour Supplement, December 27.

UKAEA, 1987, **Code of Practice and Guidance Note: Radiological Guidelines for the Design and Operation of UKAEA Plant.** Safety and Reliability Directorate Report, SRD R 456.

USNRC, 1985, **Health Effects Model for Nuclear Power Plant Accident Consequence Analysis.** NUREG/CR-4214.

TABLE 6.1
Loss of Life Expectancy per Death for Various Fatality Risks

Cause of Death	Loss of life expectancy per death (1)
All accidents (2)	15 years
All cancers (2)	13 years
Immediate death in major disaster (3)	39 years
Delayed death (4) (nuclear accident)	14 years
Delayed death (5) (natural background)	12 years

Notes

(1) Indicative figures based on 1984 population and mortality statistics for England and Wales (OPCS, 1985) and for Scotland (RGS, 1985), and comparisons with changes to hazard rates as described in notes (2)-(5) below.

(2) Loss of life expectancy (LE) calculated as gain in LE if 'cause of death' mortality rate subtracted from all causes mortality rate for all ages.

(3) This applies to any disaster (nuclear accident, dam failure, ...) which kills uniformly across the population, and in the nuclear context is relevant to early death. It is modelled as a one-off chance of death which is equal for both sexes and all age groups of the population.

(4) Probability of death due to cancer obtained using the Harvard model (USNRC, 1985) assuming uniform dose delivered to entire population at slow dose rate. LLE calculations are indicative only and do not imply endorsement of the particular model used.

(5) Radiation cancer hazard calculated as a function of age as in Hughes and Roberts (1984) assuming continuous irradiation of the entire population at a rate of 2 mSv per year.

TABLE 6.2
Risk Elements Involved in Transformation of In-Situ Coal to Electricity : British Best Estimates
(Source: Rose and Billington, 1987)

Stage	Deaths per GW-year Electricity			
	OCCUPATIONAL		PUBLIC	
	Fatal Accident	Fatal Disease	Fatal Accident	Fatal Disease
Harvesting	1.1	0.3	≪ 0.1	0.1
Upgrading	(1)	0.0	≪ 0.1	0.0
Transport	<0.1	0.0	0.1	0.0
Generation	0.3	0.0	≪ 0.1	<0.1 or 10
Waste disposal	(2)	0.0	(3)	0.0
TOTALS	1.4	0.3	0.1	0.1 or 10
OVERALL	1.7		0.2 or 10.1	

Notes

(1) Included with values for harvesting and generation
(2) Not known, probably < 0.1
(3) Not known

TABLE 6.3
Summary of Candidate Reference Risk Levels (Tolerated Risks)

These levels will apply to different activities and will need to be modified accordingly when used as references. For competing technologies, it is the fractional contribution to electricity production; for deductions from individual risk targets, it is the fractional contribution to the national economy; for natural radiation, no modification would be required.

Source of reference	Risk category	No. effects per year, GB	Possible modifying factors (1)
Competing Technology (Coal)	Early death	2	(2)
	Late death	2 or 200	(2)
	Injury	(3)	
	Evacuation	(3)	
	Birth defects	(4)	
	Non-personal	(3)	
Tolerable Individual Risk levels	Early death)	55	
	Late death)		
	Injury	(5)	
	Evacuation	(5)	
	Birth defects	(5)	
	Non-personal	(4)	
Background Radiation	Early death	(4)	
	Late death	1500	1.2-2.4 (6)
	Injury	(4)	
	Evacuation	(4)	
	Birth defects	470	
	Non-personal	(4)	

Notes

(1) The modification factor is a number by which the risk in this table may be divided to equivalence more closely social impacts from the source in question with those which would result from nuclear accidents.

(2) Differences in ages affected suggest a modification factor should be sought to allow for LLE, but information on age at death from the coal hazards is not available.

(3) These categories of risk are associated with the source in question to some extent, but no means is available to make a reasonable comparison.

(4) Categories of risk not applicable to this source.

(5) This approach could in principle be adopted for these risk categories, but no reasonably substantiated tolerable individual risk definitions are available at present.

(6) Factors allowing for LLE differences and for voluntary components of reference risk.

CHAPTER 7

RISK TARGETS IN REGULATION

7.1 Introduction

This chapter reviews some of the approaches to the regulation of major industrial hazards which have been taken in various countries, with particular reference to the UK. The coverage is not complete; only those aspects which appear to be relevant to the problem of managing social risk in the UK are considered. We shall be particularly concerned with quantitative approaches.

General safety requirements (that is, not specific to nuclear hazards) in several EEC countries and the US are described in section 7.2. This is followed by a discussion of the development of probabilistic methods and their use for regulation in section 7.3. This sets the scene for a fuller description of nuclear regulation in several countries in sections 7.4 and 7.5.

7.2 General Safety Requirements in the EEC and the US

This section reviews the general safety requirements for several countries, including the UK. All the countries except the US are members of the European Community and subject to the directive on major industrial hazards (the so-called Seveso directive) issued by the Council of the European Community (EEC, 1982) following the Seveso disaster in 1976. The directive sets out the substances and quantities considered hazardous and for which the adequacy of the precautions taken has to be justified.

7.2.1 The UK

Industrial safety requirements in the UK have been under development for many years. The present requirements have their origins in the Health and Morals of Apprentices Act of 1802 (Samuels, 1969) and the Factory Act of 1833. It was the Factory Act of 1833 that first introduced the principle of government inspection of factories (Chicken, 1975). The concept of "as safe as reasonably practicable," which is part of the philosophy of the Health and Safety at Work Act of 1974 first came into prominence in the Edwards v National Coal Board case in 1949, when Lord Asquith stated (Asquith, 1949):

> 'Reasonably practicable' is a narrower term than 'physically possible.' It seems to me to imply that a computation must be made by the owner in which the quantum of risk is placed on one scale and the sacrifice involved in the measure necessary for averting the risk (whether in money, time or trouble) is placed in the other, and that, if it be shown that there is a gross disproportion between them - the risk being insignificant in relation to the sacrifice - the defendants discharged the onus on them. Moreover, this computation falls to be made by the owner at a point of time anterior to the accident. The questions he has to answer are, firstly, what measures are necessary to and sufficient to prevent any breach (of the statute), and secondly, are these measures reasonably practicable.

The Asquith statement is important for the fact that it drew attention to the need for the quantification of risk.

The way legal obligations are defined and compliance monitored was set out by HM Chief Inspector of Factories (HMCIF, 1968). The law places the primary obligation for safety, health and welfare with the employer and inspection has to be on a sampling basis. Enforcement by sampling cannot ensure rigid compliance all the time, but better compliance is obtained if the owner of the installation understands that compliance is a matter of good practice. The Factory Acts are also more difficult to enforce than other legislation as non-compliance is sometimes a matter of opinion rather than fact. Even with these caveats about 1200 cases were taken to court in the period from 1 April 1976 to 31 March 1977 (HSC, 1978). In general the fines were very modest, being limited by the maximum fine magistrates courts could impose which in 1977 was increased from £400 to £1000 (HSC, 1978).

A further insight into the modus operandi of the Factory Inspectorate is given by the series of essays by members of the Inspectorate to mark the 150th anniversary of its founding (HMSO, 1983). Although this document contains an exclusion clause stating that the opinions expressed are those of individual authors and not necessarily authoritative statements of the Health and Safety Executive, the essays do give an indication of the thinking of working inspectors. The three following quotations show how the role of the Inspectorate is defined and how 'acceptable' and 'safe' are seen to be defined:

> The Inspectorate's role was and continues to be fourfold:
> (a) identifying problems and hazards,
> (b) influencing legislative standards,
> (c) advising on and
> (d) enforcing those standards at the workplace.
>
> At the level of the workplace questions concerning what is 'acceptable' or 'reasonable' are more open to discussion between the inspector, the employers and the trade unions and thus if the inspector's perception of what is acceptable is to retain its influence at the workplace he must carry with him not only the employer but also the representatives of the workpeople, usually their safety representatives. In the last resort inspectors can 'impose' a viewpoint but the bulk of the inspector's achievements are secured by discussion and agreement rather than by imposition.
>
> A working platform can be 'safe' in the sense that it complies with the law or with the best trade practice or with the most intricate system of safeguards and yet its 'safety will still only be relative.'

The reason for describing at length the approach of the Factory Inspectorate to what is acceptable is that it shows how the practical flexibility of the approach of British regulatory authorities was justified in the past. Flexibility continues to play an important part in the definition of acceptability, as we shall see.

Although the British approach can be summarised in qualitative terms as requiring 'the best practical precautions' and that the risks are kept 'as low as reasonably practicable' (ALARP), it is recognised that acceptable risk should ideally be defined in quantitative terms. The difficulties associated with this are only slowly being dealt with.

The accident at Flixborough in 1974 led to the formation of the Advisory Committee on Major Hazards (ACMH) to advise the Health and Safety

Commission on the wider problems created by developing technology. Its first report (ACMH, 1976) contained some statements of a probabilistic nature:

> If for instance, such tentative conclusions indicated with reasonable confidence that in a particular plant a serious accident was unlikely to occur more often than once in 10,000 years (or - to put it another way - a 1 in 10,000 chance in any one year), this might perhaps be regarded as just on the borderline of acceptability, bearing in mind the known background of risks faced every day by the general public.

The HSC intended to implement the proposals of the ACMH, but were overtaken by events in that it was decided to await the Seveso Directive, and frame legislation accordingly. The result was the Control of Industrial Major Accident Hazards (CIMAH) Regulations 1984 (HSE, 1985). These regulations require: identification of major hazards; notification of accidents; preparation of a safety report; preparation of emergency plans; and provision of information to the public. Although the Regulations specify in some detail what the safety report should contain, it has not yet been resolved whether quantitative risk assessment is necessary. Similarly the levels of risk which might be considered tolerable or otherwise have not been discussed.

One approach to overcoming some of the difficulties with ALARA or ALARP is the categorisation of risk. The Department of the Environment and the Welsh Office (DoEWO, 1984) have explained how they expect the Health and Safety Executive to advise the local planning authorities on the acceptability of hazardous developments. For this purpose the HSE would rank their advice on the acceptability of a proposal in one of three grades: negligible risk, marginal risk and substantial risk. Negligible risk is defined as being not of sufficient significance to justify refusing planning permission. Marginal risks are very small and they are not in themselves grounds for refusing planning permission, but they could be a factor against development to add to other adverse factors. A substantial risk is defined as one for which the consequences of the proposed development could be serious and the planning authority would be advised not to grant planning permission. It is specifically stated that if other factors weighed strongly in favour of the application for planning approval the HSE would seek to explain the technical assessment and the level of risk in more detail before the local planning authority made its decision.

Ranking or categorisation has also been used to rank the degree of pollution in rivers. Four categories have been used to reflect the level of pollution. They are: Class 1 for unpolluted, Class 2 for doubtful, Class 3 for poor and Class 4 for grossly polluted (DoE, 1979).

The foregoing refers only to the regulatory position in the UK. In addition an industry may set its own standards. These standards may be expressed in more positive quantitative terms than for regulatory requirements. Examples of how industries specify requirements are the Mond Index and Dow Index used in the chemical industry and the design requirements the CEGB (1982a, 1982b) lays down for its nuclear power reactors. The latter are discussed further in Chapters 3 and 8.

7.2.2 Germany

German practices consist of a set of detailed requirements to be satisfied by potentially hazardous installations. These requirements call for a justification of the adequacy of the design, construction and operating procedures to be presented to the regulatory authority in the form of a safety analysis (Chicken, 1986). The operating requirements specifically call for precautions to prevent operator error. The specification for the safety analysis implies that the acceptability of many features of a proposal have to be justified in quantitative terms, but stops short of demanding that the assessment of the inherent risks is expressed in probability terms. There is also no attempt to identify the level of risk that would be acceptable, even though the methods of analysis specified could in practice give an assessment of the level of risk in probability terms. This suggests that the German Regulatory Authorities regard the quantitative assessment of risk as simply supplementing the Regulatory Authorities' professional judgement about the acceptability of risk.

7.2.3 France

The French outlook on the acceptability of risks has been conditioned, to some extent, by several accidents that were experienced in the late 1960s and the 1970s. The accidents included Feyzin, Levin, St Amand-les-Eaux and Pierre-Benite. The Feyzin incident in 1966 was instrumental in initiating a general revision of the French safety system and the introduction of a new law in 1976. As a result of the 1976 act a decree was made in 1977 that requires installations in which there are more than certain threshold quantities of hazardous materials to be supported by a technical justification presenting an evaluation of the hazards and the safety precautions. The technical regulations lay down the measures that must be adopted to prevent fires, explosions or accidental discharges. It is claimed that the legislation is aimed at creating open procedures which allow true expression of different points of view and, as such, ensure that the final decision will be a real arbitration - enlightened by deep technical and economic analysis (Lagadec, 1982). In 1981 the minister for the Environment announced his decision to apply these in-depth safety study techniques to the most dangerous installations: some 20 installations per year were to be studied (Chicken, 1986).

As a result of continuing EEC investigation of the application of the Seveso Directive the situation in France has been updated and clarified (Gilby, 1987). Authorisation for industrial plant is a matter for local decision by the Prefecture; thus the ministry for the environment does not have specific responsibilities, but is consulted on an interactive basis. The procedure includes a requirement for independent expert review by groups of recognised competence where the studies show aspects which are considered to be of special significance. Attitudes vary concerning the extent to which probabilistic methods should be used in the technical justification.

The use of any prescribed approach has not been accepted by the regulatory authority because of their stated need for flexibility on a case by case basis. The existence of quantified rules is therefore regarded as a disadvantage although industry maintain the view that such standards are necessary to provide a framework for the discussion of the need for additional safety measures.

As an example, we describe one method which has been developed by Total (Hubert, 1987b) and applied by other organisations. The method involves coarse estimates of probability falling into one of six categories and treats consequences likewise. The probability categories have both descriptive (rare, frequent, ...) and quantitative definitions whereas the consequence categories combine all types of harm into the following scale: minor, significant, critical, highly critical, catastrophic. Each accident sequence then falls into one of four classes depending on the combination of probability and frequency categories which it has. These classes are: not severe, severe, very severe, catastrophic. This assignment is thought to incorporate 'high consequence aversion'.

7.2.4 Denmark

In Denmark about 50 installations have been identified as falling within the scope of the Seveso Directive. For a chemical waste plant, risks of the order of 10^{-6} per year for the most exposed person becoming a fatality have been endorsed as acceptable. The unique feature of the Danish system is that the Danish regulatory authority arrive at their decision about the acceptability of the risks associated with a particular installation only after assessing the technical justification and assessing the views of the Safety Committee of the organisation involved (Chicken, 1986).

7.2.5 The Netherlands

The Dutch procedure is rather different to that adopted by the other countries in that it is rather more specific. The Dutch procedure requires a Safety Report to be prepared if an accident could seriously endanger life 100 metres away. For plant with the highest hazard potential the probability must be quantified as accurately as possible. About 60 installations will be reviewed as a consequence of the Seveso directive and by 1989 they will have to present a quantitative risk analysis of their activities.

There is still some discussion about how the acceptability of risk will be defined in the Netherlands, but the current position is set out by Versteeg and Visser (1987). Excessive, and hence unacceptable, risk levels are those where the individual risk is greater than 10^{-6} per year. For individual risks between 10^{-6} per year and 10^{-8} per year, risk reduction is desirable and the ALARA principle should be applied. Risks of 10^{-8} per year and below would be considered normal and no action would be required. The Dutch justification for this view is that they consider an industrial activity may not increase the mortality of 10-15 year old children by more than 1% and this gives the upper bound of acceptable individual risk of 10^{-6} per year.

They also currently support the view that the seriousness of a potential accident increases as the square of the number of people it would kill. This enables a societal target for early death to be constructed in CCDF form. The base point used is the frequency of accidents killing 10 or more people. This is taken as 10^{-5} per year and 10^{-7} per year for the upper and lower levels respectively; the resulting CCDF is shown in Figure 3.10. It is not clear how the figures were derived. Earlier reports contain a 'consistency' argument in which the societal goal is derived from the individual risk target. As we said in Chapter 2, such consistency arguments are not cogent; the official position is now that the two are unrelated. Nevertheless, this Dutch target remains the most highly developed societal safety goal, although it reflects a considerable degree of 'high consequence aversion'. We concluded in Chapter 6 that there is currently no satisfactory basis for incorporating such aversion to early effects (and that it should not be used for cancer risks). Moreover, the resulting frequency targets are very low, and doubts have been expressed as to whether they are realistic.

It is interesting to note the different positions taken by Dutch industry and the Dutch Government on the development and use of risk criteria. Industry in the Netherlands is strongly opposed to this, while the Government have been pressing ahead with such targets. This is the exact opposite of the position in France - see subsection 7.2.3. Indeed, as a result of a Parliamentary decision, the societal risk target described above is a factor of 10 lower than that recommended by the technical experts in the ministries.

Categorisation of hazards is also used in the Netherlands for some purposes; for example, the Fire and Explosion Index and the Toxicity Index. Three categories are identified. As a further example, a guide to the degree of gas explosion resistance electrical apparatus must have in hazardous areas is categorised in one of four zone types (DGL, 1980):

Zone 1	where an explosive atmosphere is continuously present or present for long periods,
Zone 2	where an explosive atmosphere is likely to occur in normal operations,
Zone 3	where an explosive atmosphere is not likely to occur and if it does occur will only exist for a short time,
Zone 4	is a non-hazardous zone where an explosive atmosphere is not expected to be present in quantities that would require special precautions for the construction and use of electrical apparatus.

7.2.6 The US

As we shall see the US has been in the forefront in examining quantitative nuclear safety goals so it is of interest to review what has happened in the non-nuclear area.

The main legal requirements are contained in the Toxic Catastrophe Prevention Act which requires a formal risk assessment where there is the potential for a release of five times the reportable quantity. This should involve probability and consequence analysis. There is also a formal requirement for the preparation of a risk reduction plan. The implementation of these provisions, both at Federal and State (particularly New Jersey and California) level, has been significantly speeded up as a result of accident at Bhopal.

The US chemical industry has also combined in a number of activities such as the setting up of a Center for Chemical Process Safety which has published guidelines for both hazard evaluation and risk assessment procedures. These do not include guidance on acceptable risk levels. However, among particular industrial groups in the US (for example Dupont, Dow, Monsanto, Air Products) there has been considerable development of techniques and applications. Dupont in particular has recognised the need to set objectives for levels of risk as part of aid to management in decision making. The fundamental quantity on which this process is based is the Process Hazard Index which appears to be the inverse of the expected number of deaths per year, that is, it depends on the integrated societal risk, with no 'high consequence aversion'. Priority is to be given to ensuring that it reaches 10,000 years and, further, to consider improvements which would take it considerably beyond this level.

7.2.7 International activities

Various international groups are active with the aim of harmonising national approaches to the development of industrial risk safety standards.

Within the EEC a major focus is the revision of the Seveso Directive to widen its scope in the light of practical experience of its application (Bennett, 1988). However, the Commission has judged that it would be premature to attempt to include numerical risk standards even though there are a variety of activities in the nuclear field which have this objective.

The IAEA has also sponsored activities which are not specifically nuclear, for example cost benefit studies associated with alternative means of energy product. Here again there has been limited interest in quantitative targets.

A number of activities has been undertaken by the Group of Economic Experts within the OECD. These include reviews of decision aiding techniques, and of the attitude of member states to quantitative risk standards and assessment methods. A synthesis of the results of this latter is given by Hubert (1987b). The conclusions identify major differences in national attitudes and also between members of common interest groups such as the chemical industry. Because the particular Group had a limited interest in the overall subject, the focus has now shifted to a new group on Accident Control within the Chemical Industry which was set up in 1987 with a three year programme of work.

This is only a selection of the activities which are undertaken by a number of international bodies including the WHO and the World Bank.

7.2.8 Conclusions

The way non-nuclear risk acceptability is defined in the five EEC countries considered is summarised in Table 7.1. The table shows that the role of quantification of risk as a guide to assessing acceptability is recognised. There is, however, a lack of agreement about exactly how an acceptable risk should be defined. In general the situation for individual risk is better developed than that for societal risk which has only been seriously looked at in the Netherlands. This summary suggests that in practical terms the majority of countries regard the quantitative assessment of risk simply as an aid to decision making and a supplement to the regulatory authorities' professional judgement about the acceptability of risk. In particular, this applies to the UK.

7.3 The Use of Probabilistic Concepts

This section reviews the linked history of the probabilistic methods described in Chapter 5 and the idea of quantitative risk targets. We have already seen in our discussion of non-nuclear regulation how closely the two are related. The development of PSA methods was prompted by the desire to have an engineering approach which was as quantitative as possible. Since such an approach enabled risk to be calculated, it was a simple step to consider using it for regulatory purposes which is the main interest here.

The earliest developments took place in the nuclear industry and these are described in subsection 7.3.1. Some non-nuclear applications are discussed in subsection 7.3.2. More recent events in the nuclear context are described in later sections.

7.3.1 Early developments in the nuclear industry

The idea of estimating the probabilities of undesired events began in the telecommunications field and gathered momentum with the development of reliability theory following the Second World War (Green and Bourne, 1972). This was used in the aviation and electronics industries as a way of predicting whether a system would perform its function with sufficiently high probability. It was logical to try to extend this idea from events simply representing the success or failure of components or systems to cover the whole spectrum of reactor accident sequences including offsite health, and other consequences.

In the nuclear industry the WASH-740 report (USAEC, 1957) had indicated the possibility of accidents with very severe consequences. Awareness of this was heightened by the Windscale fire in 1957 and the SL-1 accident in 1961. Accidents in which many people were killed were clearly unlikely, but to an unquantified degree.

In the UK a major step forward was taken with the publication by Farmer (1967) of a new approach to reactor siting policy. This was essentially a proposal to carry out a Level 2 PSA (see Chapter 5 and Appendix 2). The frequency of the release, in terms of iodine-131 equivalent, due to each accident sequence was to be compared with a target. The technicalities of this type of safety goal are discussed in Chapter 3; the description of the various levels of PSA is given in Chapter 5. The analysis was extended to Level 3 in the sense that Farmer estimated the risks to health assuming a typical site. He also gave some detail on how to carry out the Level 1 stage, taking AGR depressurisation as the initiating event, exploring the various possibilities using an event tree, and using a compilation of component failure data. Perhaps the major difference between Farmer's proposals in 1967 and what would now be considered good practice was the lack of emphasis on completeness.

The next seminal event in nuclear PSA was the publication of the Reactor Safety Study (RSS, WASH-1400, the Rasmussen Report; USNRC, 1975). This was the first assessment recognisable as a modern PSA and covered two specific nuclear plants in the US. Although many of the methods used in the RSS were already available, many more had to be specially developed and the production of such a comprehensive study was a major achievement. The CCDFs for early deaths, latent fatal cancers, relocation and decontamination are shown in Figures 3.6-8.

The report promoted considerable discussion about the acceptability of such methods and the confidence which could be placed in the results obtained. In 1977 Saul Levine, Project Director of the USNRC, presented a paper on the RSS to an IAEA meeting in which he drew the following conclusion about the methodology the report proposed (Levine et al, 1977):

> The growing acceptance of the methodology ensures that the future will bring more widespread use of these techniques in ways that are found to be useful. It is already clear that it can be used to supplement tools already available to licensing bodies. The extent to which additional uses will be found depends on additional development effort. However, it will be necessary to approach such developments with caution because the methodology can easily be misused. We will all have to concentrate on ways to improve and codify the methodology to help ensure its continued utilization in competent ways.

Attention is drawn to the fact that Levine saw the techniques supplementing rather than replacing existing techniques.

Discussion of the proposals made in the RSS continued and in 1977 the Chairman of the USNRC asked H W Lewis to chair an independent review of the report. In September 1978 the review group presented their report and, although it was critical of some aspects of the RSS, particularly the presentation, the summary of the findings of the review group included the following statements (USNRC, 1978):

> We do find that the methodology, which was an important advance over earlier methodologies applied to reactor risks, is sound, and should be developed and used more widely under circumstances in which there is an adequate database or sufficient technical expertise to insert credible

subjective probabilities into the calculation. Even when only bounds for certain parameters can be obtained, the method is still useful if the results are properly stated. Proper application of the methodology can therefore provide a tool for the NRC to make the licensing and regulatory process more rational, in more properly matching its resources (research, quality assurance, inspection, licensing regulations) to the risks provided by the proper application of the methodology. NRC has moved somewhat in this direction, but we recommend a faster pace.

In summary, we find that the fault-tree/event-tree methodology is sound, and both can and should be more widely used by NRC. The implementation of this methodology in Wash-1400 was a pioneering step, but leaves much to be desired.

However, the RSS, and the PSA approach generally, fell somewhat into disrepute in the period following the Lewis report. This situation was changed by the Three Mile Island (TMI) meltdown in 1979 which indicated that serious accidents in reactors were a real possibility, and that there was a need to assess the probability of such accidents and to discover their main causes. Although the RSS had not identified the precise sequence which occurred at TMI, it had found transients, small loss of coolant accidents and human error to be dominant contributors to the overall risk. All three had been factors in the TMI accident. Thus the insights obtained from the RSS were shown to be valuable, and work began on using PSA to improve and extend them.

Some of this work concentrated on the immediate post-TMI issues such as the reliability of auxiliary feedwater systems. This turned out to vary over two orders of magnitude for plants meeting the same regulatory requirements. Results such as this emphasise the importance of PSA methods for the operators of plant and attention consequently shifted away from regulatory aspects.

This stimulated a considerable amount of PSA work on US reactors in the 1980s to the extent that by 1985 there were some 20 core melt frequency estimates available (Konstantinov, 1985). The current state of the art is reflected in the Reactor Risk Reference Document (NUREG-1150) in which the USNRC (1987) attempt to explore the possibility of setting a standard for LWR PSAs. This document has led to further, continuing controversy in the US. It is described further in subsection 7.5.3.

Turning now to Britain, in 1976 the Royal Commission on Environmental Pollution published their report on Nuclear Power and the Environment which they had been considering since 1974 (RCEP, 1976). In their report they examined the Farmer approach. They drew attention to the view that: risks of death greater than about 1 in 1000 per year were considered unacceptable; for risks of order 1 in 10,000 per year the public would be willing to spend money to reduce the risk; for risks below 1 in 100,000 per year warnings would be given; and risks below one in a million per year are generally accepted without concern. Deaths due to Acts of God generally have a frequency of one in a million per year. At about the same time, Chicken (1975) justified similar upper and lower limits to risk acceptability. In the US Starr et al (1976) postulated a similar scheme for risk acceptability classifying risk limits as negligible and excessive, the acceptability varying with both risk of death and the benefits. Lord Ashby (1978) endorsed the ranking of risk acceptability in probability terms and also the Starr et al approach to allowing for the influence of the magnitude of the benefit on the level of acceptable risk.

In their report, the Royal Commission on Environmental Pollution also endorsed the ICRP whole body dose limit to the public of 5 mSv per year, and recognised that such a dose had the probability of about 1 in 20,000 per year that it may eventually cause death. This risk is close to the level recognised by the Commission as that at which the public would be willing to spend money to reduce the risk. The Commission drew attention to the following criteria that determine the acceptability of sources of radiation:

1) Irrespective of cost, no member of the public shall receive more than the relevant ICRP dose limits (for example, 5 mSv per year to the whole body).
2) Irrespective of cost, the whole population of the country shall not receive an average of more than 10 mSv per person in 30 years (that is, one-fifth of the ICRP limit relating to the genetic risk).
3) What is reasonably practical should be done, having regard to cost, convenience and the national importance of the subject, to reduce doses far below the above levels.

Finally, the Commission called for a review of the criteria and methods of work of the NII.

7.3.2 Probabilistic methods outside the nuclear industry

An indication of the extent to which the acceptability of risks has been judged in probability terms in other UK industries is given in the Moss Moran pipeline and Canvey Island development decisions. In both cases the risk was described in probability terms.

Canvey Island is a complex of industrial, chemical and petrochemical plant which was the subject of intensive risk analysis arising from three public inquiries. Although the methods used are similar to those described in Chapter 5, the two analyses (HSE, 1978; HSE, 1981) were performed using a deliberately conservative approach, though less so in the case of the second analysis. The maximum individual risk estimate was reduced from 2.6x10-3 per year to 6.7×10^{-5} per year. This reduction is attributed to the less pessimistic analysis, changes made as a result of the first report, and normal economic influences. The societal risk resulting from the second study is shown in Figure 4.10 which also shows the results of the Sizewell PSA, which, however, does not use exclusively pessimistic assumptions. The authors of an appraisal of the Canvey study from the PSA viewpoint (Barrell et al, 1985) concluded that the method:

> .....provided a powerful and valuable tool for determining the significance of the various sources of hazard, and for laying them open to scrutiny to aid decision making. Complex issues were involved, particularly the important question of the potential for interactions between installations, and the assessments provided essential information which would not otherwise have been available.

The Moss Moran liquefied natural gas terminal facilities and pipeline were planned as part of the range of facilities required to exploit the Brent oil and gas field in the North Sea. A public inquiry exposed a concern about about the possibility of a vapour cloud being ignited by radio frequency transmissions. Assessments by the HSE (1980) showed that the total interaction frequency for the pipeline fell in the range $1\text{-}4 \times 10^{-6}$ per year, these interactions being leaks from the pipeline putting people in the area at risk. The report advised:

> ...the level of risk would not be such as to lead to a recommendation that a Construction Authorisation should be withheld on health and safety grounds.

Although consultants retained by the local authorities made consequence calculations they expressed the probability of various types of failure in qualitative terms, such as 'low', 'very low' and so on and gave no estimate of the possible number of fatalities. A protest group estimated that the probability of an individual fatality was 7×10^{-4} per year. Again, it was a public inquiry which prompted the use of probabilistic techniques. This is a trend which has meant that PSA has become accepted as the norm in the UK for large developments, without any explicit requirement.

The extent to which estimates of societal risk were employed in making decisions in these UK cases is discussed further in section 8.3 where it is apparent that the ability of probabilistic methods to calculate societal risk has outstripped the ability of decision makers to use the information. On an international scale too, societal risk estimates form an integral part of safety assessments. A recent OECD survey of assessments in the transportation field (Hubert, 1987a) covered 12 such assessments and CCDF output is given in virtually all cases. A survey of these assessments shows they were carried out for a variety of purposes, using various methods. Of course it should be recognised that societal risk is a particular issue for transport hazards since material in transit will generally present a small individual risk at any given point on its route, but may still have a significant impact on society.

The Netherlands provide several interesting case histories where societal risk was an issue. In the first case, a plan to extend a plastics factory was turned down in 1980 because the increased handling of chlorine would have brought a neighbouring town into a high exposure area. The factory then agreed to make its own chlorine, thus solving the problem. In 1982 several measures were taken following an assessment of the LPG chain including: technical modifications to barges; exclusion distances based on the individual risk target mentioned in section 7.2.5; the closure of service stations too close to housing; and similar exclusion distances for gas pipelines.

In the Netherlands, the Rijnmond decision related to the choice between Maasvlakte near Rotterdam and Eemshaven in Groningen as the site of a liquefied natural gas terminal. It was recognised that siting an LNG plant involved issues such as energy policy, the environment, safety, land use and regional planning, and the decision became a Cabinet issue. The Cabinet announced its preference for Eemshaven on socio-economic and regional industrial grounds in August 1978, and this was approved by Parliament. A quantitative comparison of the risk associated with each site was carried out (Kunreuther and Linnerooth, 1983). The figures related to: the probability of a major accident; the consequences; individual risk; and an overall risk measure. Although the individual risk for Eemshaven was less than 3×10^{-7} per year compared with 3×10^{-6} per year for Maasvlakte, there was little difference in the overall risk measure.

Finally an assessment of the Dutch State Mines complex led to a proposed building ban in zones where the individual risk is over 10^{-5} per year, and replacement of existing buildings only in zones where it lies between 10^{-6} and 10^{-5} per year. However, this ban will not be put into effect until the costs of technical means of reducing the risk have been compared with those of the zoning programme.

Outside the nuclear and process industries probabilistic methods and targets are also widely used. The way they have been adopted in the aircraft industry is in many ways a model for all such applications. Civil airworthiness authorities of most countries have adopted the Joint Airworthiness Requirements as the standard of airworthiness that aircraft in their country must satisfy. Proof of an aeroplane's compliance with the requirements has to be demonstrated either by tests on the aeroplane or by calculations of equal accuracy. The requirements contain very detailed specification for the testing procedures to be used.

Figure 7.1 which is taken from Joint Airworthiness Requirement JAR 25 for large aeroplanes, shows how the acceptability requirements are defined. For a particular category of effect the probability of an event giving rise to such a category of effect should be within the range shown. The probabilities are expressed as the risk per hour in flight (illustrating the usefulness of different measures of risk to those in Chapter 4). Minor events are those which at most require emergency procedures to be invoked, while at the other extreme catastrophic events which could cause multiple deaths are events that must be extremely improbable, that is, less frequent than 10^{-9} per hour of flight. Adoption of safety requirements in probability terms does not appear to have met with much objection from the public, who, judging by the rate at which air travel increases each year, seem to be satisfied with aircraft safety.

7.4 Nuclear Regulation in the UK

This section is concerned largely with the regulation of nuclear reactors, the topic which has stimulated most recent debate. The situation is described in sub-sections 7.4.1 to 7.4.3 which cover the NII's approach to assessing safety, the Layfield report and the resulting HSE discussion document. Other parts of the nuclear cycle are briefly described in sub-section 7.4.4 before we conclude with a description of recent developments in the UKAEA, a body which operates both reactors and a variety of other facilities handling radioactive material.

7.4.1 The NII approach

The Nuclear Installations Act of 1965 required nuclear installations other than those owned by the UKAEA and government departments to be licensed. Licensable nuclear reactors have to satisfy the requirements of the Nuclear Installations Inspectorate (NII). The NII has published its Safety Assessment Principles (SAPs) (HMNII, 1979; the SAPs have recently been updated to bring the numerical values closer to those of the CEGB's criteria described below, but these changes do not materially affect the account that follows) which provides guidance to its inspectors. The principles contain a large number of detailed provisions which are not expressed directly in terms of dose and frequency. These detailed provisions refer to issues such as: the need to prepare detailed plans for dealing with an emergency; the requirement for redundancy and diversity in the reactor's protective systems; the need to take account of possible common mode failures; the need to state explicitly the relevance, accuracy and reliability of the data used; and place an upper limit on the reliability that can be claimed for a single line of protective equipment in the range corresponding to one failure per 10^3 to 10^5 demands.

The SAPs are split into three hierarchical sections of which only the first two are relevant to this present discussion. The first of these, the Fundamental Requirements, refers only to ALARP, for both frequency and consequence. The second, the Basic Principles, sets out assessment reference levels which relate maximum individual doses to frequency. For this purpose three dose bands are defined, the upper limit of the highest band being 1 Emergency Reference Level (ERL) for (any member of) the public. The significance of the ERL is that it represents a level of dose at which evacuation may be considered. A reference level of frequency for discrete fault sequences is then given for each band. These frequencies are shown on Figure 7.2. The frequency of more serious accidents should be "as remote as is reasonably practicable". Because these doses are to be calculated assuming specified adverse weather conditions (see Harbison and Kelly, 1985), these targets effectively refer to releases; such targets were discussed in sub-section 3.3.2. The signficiance of the assessment reference levels is that below them inspectors "need not embark on detailed working aimed at establishing whether further improvements would be legitimately described as reasonably practicable".

In addition to the SAPs, the CEGB (1982a, 1982b) has established its own safety design guidelines, which are intended to define the standards to be met by the supplier of any nuclear power station. This is in keeping with the fact that the law places primary responsibility for safety on the owner, and he in turn must give a clear directive to his suppliers about the standards the equipment they supply must satisfy. The design safety guidelines are discussed further in Chapter 8.

The long public inquiry into the CEGB's proposal to site a PWR at Sizewell provided a forum for the NII to explain publicly how it attempted to define acceptability of risk and the implications that resulted. Two points made at the Sizewell Inquiry made the NII position clear.

The first was made by John Locke, a previous Director General of the Health and Safety Executive (HSE, the overall safety regulatory body in the UK which includes the NII in its structure), who explained to the Inquiry the HSE's view on the limit of risk acceptability in the following way (HMNII, 1984):

> If the chances of a certain accident were estimated at above a certain level (for example one chance in 10,000 per year of operation) the HSE would regard this as unacceptable and the NII would refuse to allow a plant of such a design to be built. No such design has ever been put before the NII; and it has not been necessary therefore to define the precise point at which such a refusal would be made.

The second point was to illustrate the order of magnitude of the risk presented by a reactor designed to just satisfy the safety assessment principles and their extension to more severe and remote conditions. The NII presented an analysis of two notional cases, A and B (Harbison et al, 1985; Harbison and Kelly, 1985). These two cases are shown in Figure 3.9; Case B is considered to be a better representation of the NII's principles. The estimated societal risk in terms of fatal cancers for cases A and B is shown in Figure 7.3. Some of the uncertainty in the analysis is indicated by the width of the shaded areas. The conclusion that was drawn from the analysis was (Harbison et al 1985; Chicken and Harbison, 1986):

> .....that a reactor designed, constructed and operated in conformity with the NII safety assessment principles will result in a low level of risk to the most exposed individual and to the population as a whole. Moreover, because of a number of cautious assumptions adopted in the analysis, the predicted risks are likely to exceed those which might occur in practice.

In summary, the NII's approach to the problem of acceptability of risks from nuclear stations assumes that there is some upper level of risk expressed in quantified terms above which a station should not be given a licence, and below which the Inspectorate's Safety Assessment Principles should be applied. The overall objective of these Principles is to ensure that the risks from nuclear power stations are as low as reasonably practicable and preliminary work indicates that a reactor designed, constructed and operated in conformity with the Principles should present an acceptably low level of risk to exposed members of the public and society as a whole.

7.4.2 The Layfield Report

The Public Inquiry, conducted by Sir Frank Layfield, into the acceptability of siting a PWR at Sizewell subjected the British regulatory authorities' procedures to very detailed examination. In his report, Sir Frank was very critical of: the guidance given on the criteria to be used for judging acceptability; the ALARP criterion; the significance given to the consideration of human factors; the use of probabilistic methods for estimating risk; and the acceptable level of risk. The following quotations from the report encapsulate the essential messages on these topics Layfield, 1987):

> Paragraph 2.26 Probabilistic risk estimation can offer the public a useful measure of the potential risk from the plant. But the technique is at an early stage. Effort needs to be put in to improving the presentation of the results so that their use and limitations can be properly understood. I believe that if this is successfully done, the use of probabilistic risk estimation could and should become increasingly widespread within the nuclear industry and elsewhere.
>
> Paragraph 2.51 I conclude that: (a) the most meticulous and exhaustive attention must be paid to minimising the occurrence and effects of human errors. The risk from human error might otherwise exceed other risks from Sizewell B.
>
> Paragraph 2.91 There are important weaknesses in the ALARP principle as understood and applied to nuclear safety in the UK. They include:
>
> a) an absence of clear definition or widely understood meaning of ALARP. The licence applicant often did not know what was expected of it, which could vary depending on the NII Inspector concerned. Such inconsistency potentially leads to misallocation of resources, misunderstanding and confusion, and could mean that some aspects of the design may not be as safe as they reasonably should be;
> b) the relationship between ALARP and some of the NII's criteria is not clear;
> c) the choice of a multiplier to reflect gross disproportion is to some extent arbitrary, even where such a choice is consciously made;
> d) ALARP can lead to an inexorable increase in safety standards, whether or not such an increase is justified in a particular case. As a result, national resources may be misallocated between nuclear safety, and other expenditure in the private and public sectors, and the economics of nuclear power may be unreasonably handicapped.
>
> Paragraph 2.98 There was no authoritative guidance from the Government or Parliament on what the level of tolerable risk from a new nuclear power station might be, nor on how it should be determined.
>
> Paragraph 2.99 In the absence of authoritative guidance, I have had to reach a judgement on the basis of the relevant evidence from the CEGB and other parties. I conclude that a level of individual risk of death of the order of once in a million years is likely to be broadly tolerable if justified by associated benefits. On this basis, the CEGB's target of restricting the probability of an uncontrolled release of the radioactive materials to once in a million years, and its targets for design basis accidents, are reasonable.

Paragraph 2.102 As a first step in formulating guidance on tolerable levels of risk and ALARP, the HSE should publish a consultative document to enable public, expert and Parliamentary opinions to be expressed.

These conclusions from the Sizewell 'B' Public Inquiry endorse the view that further development of NII regulatory procedures and risk acceptability criteria is required. Also care must be taken that spending on risk reduction is allocated in as efficient a way as possible.

7.4.3 The HSE discussion document

The HSE (1988) published their response to Layfield's recommendation quoted above that they should formulate and publish guidelines on the tolerable levels of individual and social risks to workers and the public from nuclear power stations. The HSE document is intended to be a straightforward account written for the general public and for this reason contains a great deal of material which prepares the ground for the discussion of risk levels in which the paper culminates.

This preliminary material includes a basic discussion of risk, an account of how general industrial risk is regulated in the UK, how the HSE and NII are structured, licensing procedures for nuclear installations, how risk is assessed, the harm from radiation, risk from routine operation, the Safety Assessment Principles and the Design Safety Criteria, the nature of reactor accident assessment and emergency procedures. One significant feature is the explanation given of how ALARP is assessed. This embellishes Asquith's interpretation quoted in sub-section 7.2.1 by setting a limit of tolerability independent of benefit. Below this limit cost may be taken into account when further safety improvements are considered on a sliding scale: the higher the risk, the more, proportionately, should be spent on reducing it. And there may be a level of risk below which it is not worth spending anything at all. No specific guidance is given on the numerical value of the relationship between the cost of safety improvements and the benefit which derives from them although a discussion is given which parallels the one in the present report. This was a matter which particularly concerned Layfield.

Turning now to tolerable levels of risk, the situation is summarised in Table 7.2. The levels of individual risk are derived starting from the occupational risk statistics given in Table 4.5. This gives 10^{-3} per year as the limit of tolerability for worker risk. The HSE then argue that the corresponding figure for members of the public should be not less than ten times lower, giving 10^{-4} per year. Then, a 'broadly acceptable' level of risk to members of the public is taken to be 10^{-6} per year. The HSE consider that this "is the level of risk below which, so long as precautions are maintained, it would not be reasonable to insist on expensive further improvements to standards". Again this is derived by comparison with existing risk statistics in that it would involve a very small addition to the ordinary risks of life.

Turning now to social risk, the HSE seek to define the limit of tolerability in terms of the frequency of a significant nuclear accident anywhere in the UK. By comparison with the estimated frequency of accidents causing more than 1500 casualties at the Canvey Island complex, the estimated frequency of aircraft crashes killing more than 500 people and the design specification for overtopping the Thames Barrier, the HSE arrive at a figure of 10^{-4} per year. They then define such an accident in terms of releases in the same way as is done in the Safety Assessment Principles. Thus a major civil nuclear accident is one giving rise to an uncontrolled release of a size capable of giving doses of 100 mSv at 3 km. In terms of consequence, the HSE pessimistically estimate that an accident of this size might cause the eventual deaths from cancer of about 100 people. Such accidents are therefore less serious than the three used as comparisons "to take some account of the strong aversion felt in our society to risks from radiation". However, it should be made clear that the HSE accident is defined in terms of release, and not consequences. Thus it is independent of site characteristics such as population density, and collective effects are not controlled explicitly by this approach.

The HSE do not give details of what weather conditions are to be assumed in deciding what a release is 'capable of' nor what pessimistic assumptions about weather and surrounding population would lead to 100 cancer deaths. Thus it is not clear what size of accident is regarded as significant, though it would certainly be such as to activate emergency procedures. However, there is the possibility of nuclear reactor accidents considerably more serious than that defined by the HSE.

The HSE then develop the argument that reactors designed in accordance with the Safety Assessment Principles and the CEGB's Design Safety Criteria pose tolerable risks when judged against the levels shown in Table 7.2. With regard to individual risk to members of the public they conclude "that most people in the vicinity are at the 1 in 1 million level or better, ... but that a few are nearer to 1 in 100,000 and a handful may exceed that level". For social risk, the chance of a release of the defined size is "probably near to about 1 in 1 million. If therefore there were ever as many as 20 modern power reactors in the UK, the risk of the major accident defined above would be of the order of 1 in 50,000 per annum".

Finally, it is worth noting that the HSE say it is not for the regulatory authorities but for the public to weigh the benefits of nuclear power with the risks.

7.4.4 Other parts of the fuel cycle

The Department of the Environment, and the other authorising departments, have set a primary target for individual risk for application in the waste disposal field (DoE, 1984):

> The appropriate target applicable to a single repository at any time is, therefore, a risk to an individual in a year equivalent to that associated with a dose of 0.1 mSv: about 1 chance in a million.

Note that although this is a risk target, it is defined in terms of dose, and would thus increase with any increase in the dose-risk relationship.

Any application to construct a disposal facility will be assessed independently against this primary target. Further documents from the Departments (DoE, 1986) show that further estimates of other health, financial and environmental impacts may be made, and evaluated using a aggregation method which is considered to reflect various public perceptions. These impacts include collective doses to the workforce, local population, the regional population and the global population weighted in different ways and for different periods in the future.

Thompson (1987) compares the predicted annual individual risk as a function of time against the above risk target, which is shown as a constant for thousands of years after the closure of the facility as implied by the statement quoted above.

BNFL have anticipated the development of targets for more severe accidents by regulatory authorities through the development of a target related to individual risk. Ball and Curtis (1983) specify not only a total target of 10^{-6} per year for the risk of fatal cancer but also describe a number of rules about how the target is to be interpreted. The relationship between these rules and the way the Safety Assessment Principles (for Nuclear Chemical Plants (HMNII, 1983)) are applied was examined at the EDRP Public Inquiry as was the level of societal risk which is implied by this individual risk target for the Dounreay site.

When targets such as these are applied to nuclear chemical sites, the problem arises of apportioning the risk between various plants, and perhaps between various accident sequences (see, for example, Worswick et al, 1987). This may pose a considerable problem for the risk manager, but various straightforward conservative methods to tackle it can be used, or, if necessary, a more sophisticated whole-site calculation carried out. This will not be discussed further.

7.4.5 The UKAEA

The UKAEA operates various prototype and research reactors, as well as different types of nuclear chemical and active handling plant. Its sites were not licensed under the Nuclear Installations Acts until 1990 and it had therefore evolved its own methods of ensuring the safe operation of its plant. Farmer's early work in the development of risk estimation and evaluation techniques has already been mentioned. More recently the UKAEA issued, a Code of Practice and a Guidance Note on the way the Authority's statutory requirements with regard to radiation doses to workers and the general public should be satisfied (UKAEA, 1987). The Code of Practice sets out: routine exposure dose targets for workers and the general public for design and operation; cost benefit procedures for routine exposures; and accident individual risk targets for workers and the public. The question of societal risk is not addressed. The Code and Guidance Note are reflected in the UKAEA's current safety policy and safety directives (AEA Technology, 1991).

In respect of accidents, the Code states that the risk of death (early or delayed) to workers inside the plant should not exceed 10^{-5} per year and, for typical members of the most exposed group of the general public off-site, should not exceed 10^{-6} per year.

In the application of cost benefit analysis proposed for assessing the acceptability of expenditure for dose level reduction there is a certain flexibility left in the cost per man-sievert figure that should be used. Thus proposals costing less than £3000 per manSv should definitely be implemented, whereas proposals with costs up to £30,000 per manSv require a management decision which should take account of factors such as compliance with the other targets, individual doses, collective dose trends and social and economic considerations. More expensive proposals should not be carried out unless there are overriding considerations as detailed in the Code of Practice. The cost benefit proposals are in line with those put forward by the ICRP and the NRPB. This topic is discussed further in Appendix 1. (Note: the figures of £3,000 and £30,000 in the Code of Practice had been updated to £5,000 and £50,000 at the date of going to press in 1991).

7.5 Nuclear Regulation Elsewhere

Summaries of nuclear regulatory practice have been prepared by the NEA (1983) and the EEC (1983). From these surveys it is possible to see the wide variation in the types of target which are used in the various countries. The possibility of these many types was introduced in subsection 3.3.2. This is illustrated in Table 7.3 which sets out the different possible types of target for nuclear reactors (eight in all) and the countries in which they are used. Of course there are differences in the extent to which there is formal inclusion in national regulations. We shall not describe these in great detail, but instead choose a number of key countries for which the position is briefly described.

We note that in the case of the UK, the assessment reference levels of the NII and the CEGB frequency limits for beyond design basis accidents are marked in Table 7.3. The individual risk targets of the UKAEA and BNFL, and the social risk targets which could be inferred from the HSE consultative document are not so marked.

7.5.1 France

Unlike other OECD countries, there is in France no general law for nuclear installations (Lagadec, 1982). The many provisions that have to be satisfied have been proposed by the regulatory authority and established by a government decree. Although nuclear installations are specifically excluded from the laws governing other potentially hazardous installations they are similar in pattern. The essential steps in the licensing procedure are as follows (NEA, 1980):

(1) Application for a licence is filed with the Minister of Industry. The Minister informs the other ministries concerned, initiates the local inquiry procedure and a technical study. (Analysis of the safety implications is made by the CEA Institute for Protection and Nuclear Safety.) The local inquiry procedure is under the direction of the prefect of the department concerned and for a nuclear reactor the procedure often takes the form of a public inquiry. This process takes 6 to 12 months.

(2) The findings of the local inquiry procedure and the technical study are presented to the Minister of Industry, and if there are no outstanding obstacles he drafts a decree authorising the installation to be set up. The draft decree is then sent to the interministerial committee for large nuclear installations; this committee must give its opinion within two months. If no objections are raised a licence for the start of the project is given. This process takes 12 to 18 months from the completion of the local inquiry procedure and the analysis of the safety implications.

(3) Once the licence is published, announcing all the conditions that must be satisfied, construction of the reactor can start. Then, six months prior to loading fuel, the operator must send the Minister for Industry a preliminary safety report together with proposals for the operating instructions. In the light of these documents, and the opinion of the Interministerial Committee, the Minister will decide if loading and power testing can start. If power loading is approved the Minister will lay down a timetable for submission of the final safety report and final operating instructions. Generally these documents are required six months before approval to start normal operation can be given.

It is a feature of this procedure that the timescale for the various steps is known in advance.

So far as quantitative targets are concerned, it can be seen that France has more entries on Table 7.3 than the other countries. This development stems from EdF (the French utility) and has been accepted by the licensing authority.

Here, as in other countries, design basis radiological dose targets are supplemented by additional criteria. Thus 10^{-6} per year is specified as a non-mandatory objective for unacceptable radiological consequences. This latter term is not defined precisely but mainly depends on the extent of a feasible emergency plan for the particular site. Three reference source terms, S1, S2 and S3 have been defined for use in severe accident considerations. The most serious of these, S1, is used to cover residual risk, whereas S3 reflects the feasible emergency plan. The intermediate case S2 is the target for special procedures featured in the French planned response to major accidents.

7.5.2 Germany

As can be seen from Table 7.3, very little use is made in Germany of quantitative targets in licensing. This is similar to the non-nuclear case described earlier in subsection 7.2.2. Although the German Risk Study followed hard on the heels of the RSS in the US, it is still a subject of controversial discussion as to whether quantitative safety goals are needed. Such figures as have been produced are described as "for orientation purposes", for example, 10^{-6} per year for the non-availability of core cooling event group, and 10^{-5} per year for core melt frequency.

Papers have appeared in the literature (Gottschalk et al, 1984) in which the possibility of dose related targets (CCDFs of dose) has been discussed, though there is no current intention to include this approach in German licensing. These levels are apparently similar to the assessment reference levels of the NII, and the design basis levels used by the French as shown in Figure 7.2. However, it needs to be borne in mind that the way in which each is applied (for example, individual accident sequences by the NII and as a CCDF by the Germans, the assumptions about weather and so on) means there can be considerable differences, especially in the beyond design basis region.

7.5.3 Italy

Italy has a set of dose levels and associated frequencies for design basis accidents which is typical of those in other countries. In addition there is a frequency target for core degradation conservatively assumed to occur when design basis fuel conditions are exceeded. This has been fixed at 10^{-5} per year but with the intention of achieving 10^{-6} per year through alternative design options.

Italy is unique in having a containment performance target which is being applied prior to formal inclusion in criteria. For core melt accidents there is to be a 95% chance of not exceeding releases of iodine and caesium greater than 0.1% of the core inventory. The Italian authorities are opposed to any efforts to deduce corresponding acceptable risk levels from these targets. Zaffiro and Valeri (1987) illustrate the application of these targets to Italian LWR plants.

7.5.4 The US

The RSS stimulated interest in the use of quantitative safety goals in the US. After some exploratory analysis the USNRC (1983) issued a set of draft safety goals. Two of these were qualitative:

> Individual members of the public should be provided a level of protection from the consequences of nuclear power plant operation such that individuals bear no significant risk to life and health.
>
> Societal risks to life and health from the operation of nuclear power plants should be comparable to or less than the risks of generating electricity by viable competing technologies and should not be a significant addition to other societal risks.

These were to be implemented in terms of three quantitative goals relating to the average individual risk of early death within a radius of 1 mile, the average individual of cancer death within 50 miles and the frequency of core melt. The 1983 statement also provided for a cost benefit approach (using $100,000 per manSv) to be employed if the quantitative targets were not all met.

After a lengthy consultation process revised safety goals were announced (USNRC, 1986). The two qualitative goals were unchanged, but augmented by a third relating to core damage. The corresponding quantitative goal was removed, and replaced by a "general performance guideline" relating to the frequency of large releases. The remaining two quantitative goals were rephrased:

> The risk to an average individual in the vicinity (1 mile radius) of a nuclear power plant of prompt fatalities that might result from reactor accidents should not exceed one tenth of one per cent of the sum of prompt fatality risks resulting from other accidents to which members of the US population are generally exposed.
>
> The risk to the population in the area (10 mile radius) of a nuclear power plant of cancer fatalities that might result from nuclear power plant operation should not exceed one tenth of one per cent of the sum of cancer fatality risks resulting from all other causes.

The provision for cost benefit analysis was removed. The use of these goals is approved for regulatory decision making, though specific guidelines on this have yet to be issued.

It should be noted that, although these goals have been marked on Table 7.3 as referring to individual and societal risk, the use of average individual risk in both these quantitative targets is at variance both with the concept of maximum individual risk and that of total societal risk. The average individual risk of early death within 1 mile may be much less than the maximum; and since the total cancer risk from accidents within 10 miles is to be compared with that from other causes, no distinction between densely and sparsely populated sites is made and the total loss to society is not addressed.

In parallel with its deliberations on safety goals the USNRC has been involved in developing probabilistic methods for their application. The most recent outcome of this effort is the publication for comment of the Reactor Risk Reference Document, NUREG-1150 (USNRC, 1987), which is a compilation of PSAs carried out for six reactors using uniform methods as far as possible. The material in NUREG-1150 has a variety of uses, such as describing the state of the art in PSA methods and illustrating the uncertainty associated with PSA as well as evaluating whether the plants satisfy the Safety Goals. The overall results for core damage frequency and the performance of the plants against the two USNRC Safety Goals are shown in Figures 7.4-6. The bars on these figures represent the uncertainty using methods developed by the USNRC. It appears that most US LWRs have no difficulty in meeting the two quantitative safety goals.

7.6 Conclusions

(1) As yet there is no universal agreement about the definition of tolerable risk targets and more development is required before they reach a definitive stage.

(2) There is general unease about applying the ALARP principle. Clear guidance is needed on the precise conditions to be satisfied. The ultimate

decision has to take into account many factors, but the process by which this is done should be open and consistent. The French procedure for assessing the acceptability of major hazardous installations may contain useful elements, particularly the fixing of timescales.

(3) While there is considerable agreement that for the general public an individual risk of death of 10^{-6} per year is tolerable, there is lack of agreement about whether conditions that cause multiple deaths should be treated proportionately more restrictively than those that can cause a single death.

(4) There is general support for the use of probabilistic methods for assessing safety, but less agreement about whether 'risk' is the appropriate form for targets.

(5) Economic arguments have to be taken into account in making decisions about acceptability, but there is no universal agreement about how such assessments should be made. The UKAEA provisional Code of Practice shows how a simplified form of cost benefit analysis can be used to assess the acceptability of schemes for the reduction of radiation dose. This is discussed further in Appendix 1.

(6) No clear rationale for setting quantitative societal safety goals emerges from this international survey.

7.7 References

ACMH, 1976, **First Report of the Advisory Committee on Major Hazards.** HMSO, London.

Ashby, E, 1978, **Reconciling Man with the Environment.** Oxford University Press.

Asquith, Lord, 1949. In Edwards v National Coal Board (1949) 1 KB; (1949) 1 AII ER 743, p712 and p747, a case on the interpretation of S 102 (8) of the Coal Mines Act 1911.

Ball, P W and Curtis, L M, 1983, **Safety Criteria for Nuclear Chemical Plant.** Symposium on Safety Criteria - Requirement, Establishment and Application, 29 September 1983, Southport. Safety and Reliability Society.

Barrell, A C, Edmondson, J N, and Holden, P L, 1985, **Canvey Island - A Case Study of the Application of PRA.** IAEA Seminar on the Implications of PRA, Blackpool, 18-22 March, 1985.

Bennett, E, 1988, **Policy and Control within the European Chemical Industry.** IChemE Conference, London.

CAA, **Joint Airworthiness Requirement JAR 25.** Civil Aviation Authority.

CEGB, 1982a, **Design Safety Criteria for CEGB Nuclear Power Stations.** HS/R167/81 (Revised).

CEGB, 1982b, **PWR Design Safety Guidelines.** CEGB Generation, Development and Construction Division (DSG-2, Issue A).

Chicken, J C, 1975, **Hazard Control Policy in Britain.** Pergamon, Oxford.

Chicken, J C, 1986, **Risk Assessment for Hazardous Installations.** Pergamon, Oxford.

Chicken, J C, and Harbison, S, 1986, **Differences Between Industries in the Definition of Acceptable Risk.** Paper presented at Society for Risk Analysis Conference, Boston, November 1986.

DGL, 1980, **Guide for the Classification of Hazardous Areas in Zones in Relation to Gas Explosion Hazard and to the Installation and Selection of Electrical Apparatus.** R no 2E, Report by the Director General of Labour, Ministry of Social Affairs, Voorburg, The Netherlands.

DoE, 1979, **Second Biannual Report of the Standing Technical Advisory Committee on Water Standards.** HMSO, LONDON.

DoE, 1984, **Disposal Facilities on Land for Low and Intermediate Level Radioactive Wastes: Principles for the Protection of the Human Environment.** HMSO, London.

DoE, 1986, **Assessment of the Best Practicable Environmental Options for the management of Low and Intermediate Level Solid Radioactive Wastes.** HMSO, London.

DoEWO, 1984, **Planning Controls over Hazardous Developments.** Circular 984 issued by the Department of the Environment and the Welsh Office, HMSO, London.

EEC, 1982, **Major Accident Hazards of Certain Industrial Activities.** Council Directive of 24 June 1982 (85/501/EEC). Official Journal of the European Community 5/8/85, L230/1 Vol 25.

EEC , 1983, **Status Report on Safety Goals/Objectives.** XII/476/83 (EN), May, 1983.

Farmer, F R, 1967, **Siting Criteria - A New Approach.** IAEA Symposium on the Containment and Siting of Nuclear Power Reactors, Vienna 3-7 April, 1967. IAEA SM-89/34.

Gilby, E V, 1987, **Case Study on the Accident Risk Management of the Storage of hazardous Materials.** OECD Report ENV/ECO/87.6.

Gottschalk, P A, et al, 1984, **Konzept zur Unterstützung der Sicherheitsbeurteilung von Kernkraftwerken mit probabilistischen Methoden.** GRS report GRS-A-991, BMU 1986-133.

Green, A E, and Bourne, A J, 1972, **Reliability Technology.** Wiley.

Harbison, S A, and Kelly, G N, 1985, **An Interpretation of the Nuclear Inspectorate's Safety Assessment Principles for Accidental Releases.** IAEA Seminar on 'The Implications of PRA,' Blackpool, UK, March 1985. IAEA-SR-111/20.

Harbison, S, Kelly, G N, and Hemming, C R, 1985, **Procedures to Relate the NII Safety Assessment Principles for Nuclear Reactors to Risk.** NRPB - R180.

HMCIF, 1968, **Annual Report of HM Chief Inspector of Factories.** Cmnd 4461, HMSO, London.

HMNII, 1979, **Safety Assessment Principles for Nuclear Power Reactors.** HSE, HMSO, London.

HMNII, 1983, **Safety Assessment Principles for Nuclear Chemical Plant.** HMSO, London.

HMNII, 1984, **The Range of Risks from a PWR at Sizewell - An Overview.** Evidence presented at the Sizewell 'B' Public Inquiry, NII/S/92 (SAF).

HMSO, 1983, **Her Majesty's Inspectors of Factories 1833-1983.** HMSO, London.

HSC, 1978, **Health and Safety Commission Report 1976-77.** HMSO, London.

HSE, 1978, **Canvey, An Investigation.** HMSO, London.

HSE, 1980, **A Reappraisal of the HSE Safety Evaluation of the Proposed St Fergus to Moss Moran LNG Pipeline.** HMSO, London.

HSE, 1981, **Canvey, A Second Report, A Review of Potential Hazards from Operations in the Canvey Island/Thurrock Area Three Years after the Publication the Canvey Report.** HMSO, London.

HSE, 1985, **A Guide to the Control of Industrial Major Hazard Regulations 1984.** Health and safety series booklet, HS(R) 21, HMSO, London.

HSE, 1988, **The Tolerability of Risks from Nuclear Power Stations.** HMSO, London.

Hubert, P, 1987a, **Case Study on the Risk Assessment and Management for the Transportationof Hazardous Materials.** OECD Report ENV/ECO/87.5

Hubert, P, 1987b, **Risk Assessment and Risk Management for Accidents Connected with Industrial Activities.** OECD Report ENV/ECO/87.11.

Konstantinov, L, 1985, **PSA in Nuclear Safety: International Developments.** IAEA Seminar on the Implications of PRA, Blackpool, UK, 18-22 March, 1985. IAEA-SR-111/1.

Kunreuther, H C, and Linnerooth, J, 1983, **Risk Analysis and Decision Making.** Springer Verlag, Berlin.

Lagadec, P, 1982, **Major Technological Risk.** Pergamon, Oxford.

Layfield, Sir Frank, 1987, **Sizewell 'B' Public Inquiry Report.** HMSO, London.

Levine, S, **et al** ,1977, **The Rasmussen Report: A Retrospective and Prospective View.** IAEA Conference on Nuclear Power and its Fuel Cycle, Salzburg, Austria, 2-13 May 1977.

NEA, 1980, **Description of Licensing Systems and Inspection of Nuclear Installations.** NEA/OECD, Paris.

NEA, 1983, **Nuclear Legislation - Analytical Study.** NEA/OECD, Paris.

RCEP, 1976, **Nuclear Power and the Environment.** Cmnd 6618 Royal Commission on Environmental Pollution, 6th Report, HMSO, London.

Samuels, M, 1969, **Factory Law.** 8th edition, Charles Knight, London.

Starr, C, Rudman, R, and Whipple, C, 1976, **Philosophical Basis of Risk Analysis.** Annual Review of Energy, 1, p 630.

Thompson, B G J, 1987, **Development, by Means of Trial Assessments, of a Procedure for Radiological Risk Assessment of Underground Disposal of Low and Intermediate,Level Radioactive Wastes.** PSA '87, International Topical Conference on Probabilistic Safety Assessment and Risk Management, 30 August - 4 September 1987, Zurich, TUV Verlag.

UKAEA, 1987, **Code of Practice and Guidance Note: Radiological Guidelines for the Design and Operation of UKAEA Plant.** Safety and Reliability Directorate Report, SRD R 456.

USAEC, 1957, **Theoretical Possibilities and Consequences in Large Nuclear Power Reactors.** Wash-740.

USNRC, 1975, **Reactor Safety Study: An Assessment of Risks in US Nuclear Power Plants.** Wash-1400, NUREG-75/14.

USNRC, 1978, **Risk Assessment Review Group report to the USNRC.** NUREG/CR-0400.

USNRC, 1983, **Policy Statement on Safety Goals for the Operation of Nuclear Power Plants.** Federal Register, 48(50): 10772-10781.

USNRC, 1986, **NRC Adopts Policy Statement on Safety Goals.** Nuclear Safety, 27, No 4, pp 555-556.

USNRC, 1987, **Reactor Risk Reference Document.** NUREG-1150, Draft for Comment.

Versteeg, M F, and Visser, B J, 1987, **A PRA Guide for the Netherlands; a Consequence of the Dutch Policy on the Risk Management Applied to Nuclear Energy.** PSA '87, International Topical Conference on Probabilistic Safety Assessment and Risk Management, 30 August - 4 September 1987, Zurich, TUV Verlag.

Worswick, J T, et al, 1987. **An Assessment of the Frequency of Criticality Incidents in the Pulsed Column of a Reprocessing Plant.** PSA '87, International Topical Conference on Probabilistic Safety Assessment and Risk Management, 30 August - 4 September 1987, Zurich, TUV Verlag.

Zaffiro, C and Valeri, A, **Probabilistic Studies for the Safety Assessment of Italian NPPs.** PSA '87, International Topical Conference on Probabilistic Safety Assessment and Risk Management, 30 August - 4 September 1987, Zurich, TUV Verlag.

TABLE 7.1
Tolerability Summary of Risk Acceptability Criteria in Five Countries

COUNTRY	REGULATORY DEFINITION OF ACCEPTABILITY	OUTLINE OF JUSTIFICATION REQUIRED	USE MADE OF RISK QUANTIFICATION
BRITAIN	Risks should be as low as reasonably practicable.	A report on the activity as defined by the CIMAH regulations. In some cases acceptability may be subject to Public Inquiry.	Suggested risk of a serious accident of 10^{-4} per year on the borderline of acceptability.
GERMANY	Must satisfy technical rules and not cause harm to the enivronment or considerable dis-advantage or nuisance of the public.	Safety analysis to the latest state of technology.	Only as part of the safety analysis. No quantified targets to be satisfied are specified.
FRANCE	Real arbitration enlightened by deep technical and economic analysis.	Technical risk assessment and economic analysis.	Risk of unacceptable consequences should not exceed 10^{-6} per year is regarded as a guide rather than mandatory.
DENMARK	Requirements expressed in general terms. Environmental contamination not above threshold values.	Must be acceptable to the Safety Committee of the organisation concerned.	A risk of the most exposed person becoming a fatality of the order of 10^{-6} per year is acceptable
NETHER-LANDS	Hazard must be quantified as accurately as possible.	Safety report must be approved by regulatory authorities and Works Council. Suitability of operating staff must be assessed.	Analysis in terms of probability. Provisional maximum acceptable individual risk of death 10^{-6} per year.

TABLE 7.2
Summary of HSE Tolerable Risk Levels

	Just Tolerable	Broadly Acceptable(1)
Workers	10^{-3} per year	
Public	10^{-4} per year	10^{-6} per year
Major accident(2)	10^{-4} per year	

Notes

(1) The level of risk below which, so long as precautions are maintained, it would not be reasonable to insist on expensive further improvements to standards.

(2) A significant nuclear accident anywhere in the UK, defined as one giving rise to an uncontrolled release of a size capable of giving doses of 100 mSv at 3km.

TABLE 7.3
Quantitative Targets in Different Countries

	France	Germany	Italy	N'lands	Belgium	Spain	Sweden	UK	USA	Finland
Beyond Design Basis Frequency Limit	X							X	X	
Design Basis Dose or Release Limit	X	X	X		X	X	X			X
Design Basis Dose or Release Frequency	X		X	X	X	X		X	X	X
Beyond Design Basis Dose Release Frequency	X		X							
Core Damage Frequency	X		X							
Containment Performance			X							
Individual risk				X					X	
Societal Risk				X					X	

Notes

Frequency Limits:	frequency limits for releases expressed in genral terms, for example 'uncontrolled release'
Dose or Release Limits:	dose or release limits for specified accident classes
Dose or Release Frequency:	limiting frequencies of faults giving rise to specified doses or releases
Containment Performance:	criteria which link conditional probability to extent of failure, for example release

FIGURE 7.1
Relationship between Probability and Severity of Effects for Large Civil Aircraft
(Source: CAA)

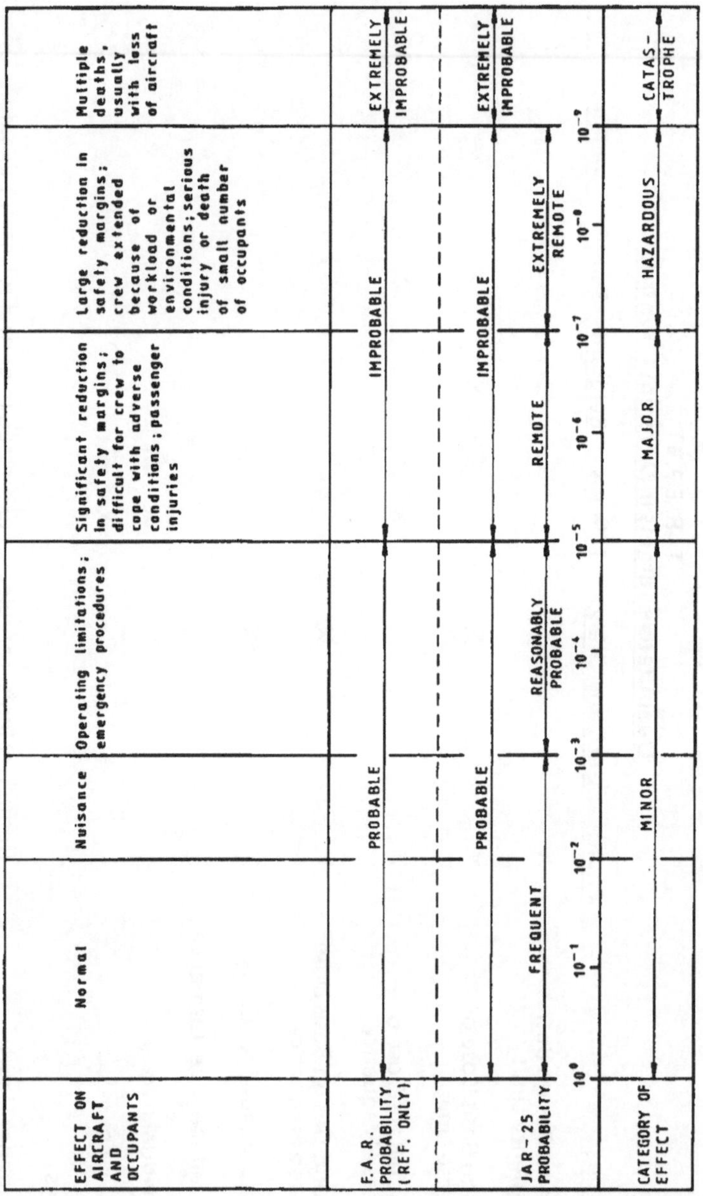

FIGURE 7.2
Target Values Being Developed for Nuclear Installations
(Source: Chicken and Harbison, 1986)

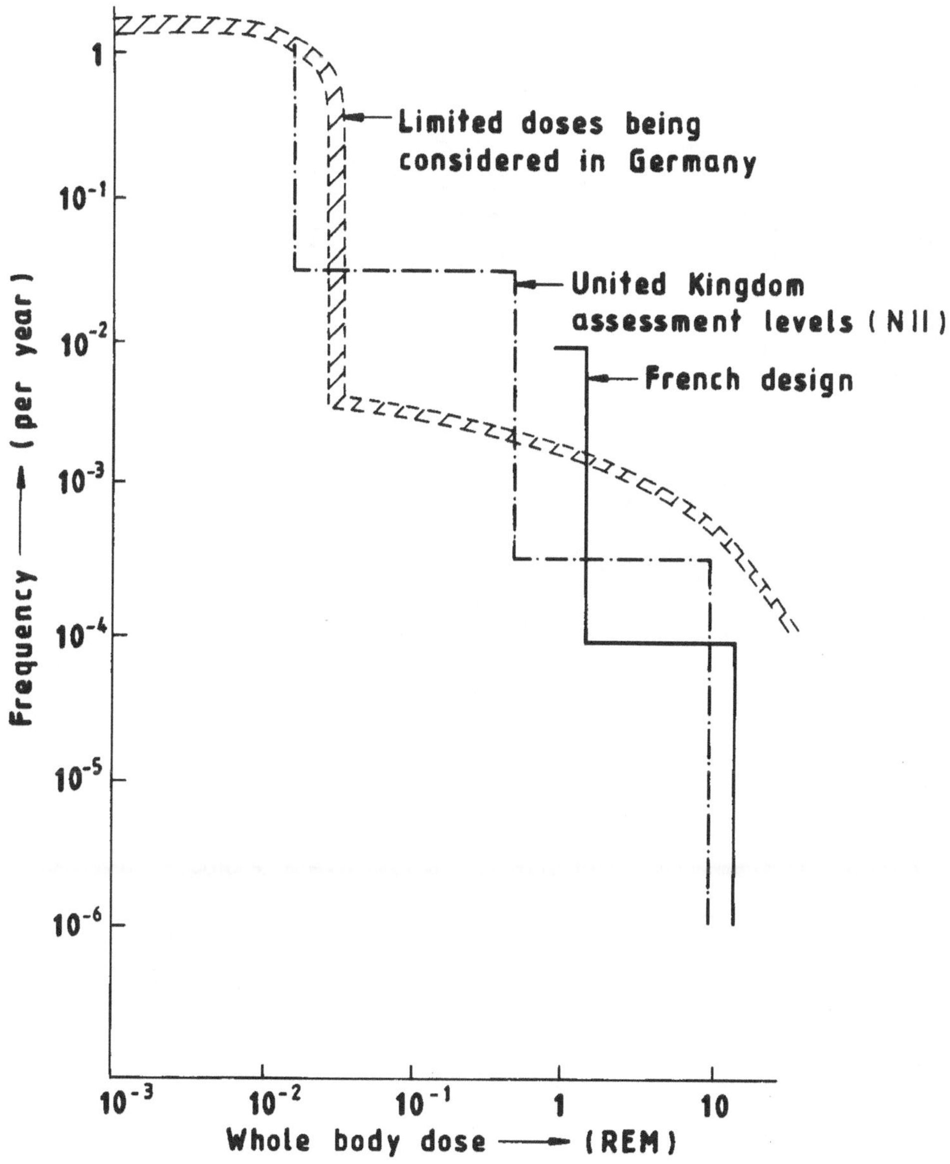

FIGURE 7.3
Societal Risk Consequent Upon a Reactor Sited at Sizewell and Designed Just to Meet the NII Principle for Accidents (and their Assumed Extension)
(Source: Harbison and Kelly, 1985)

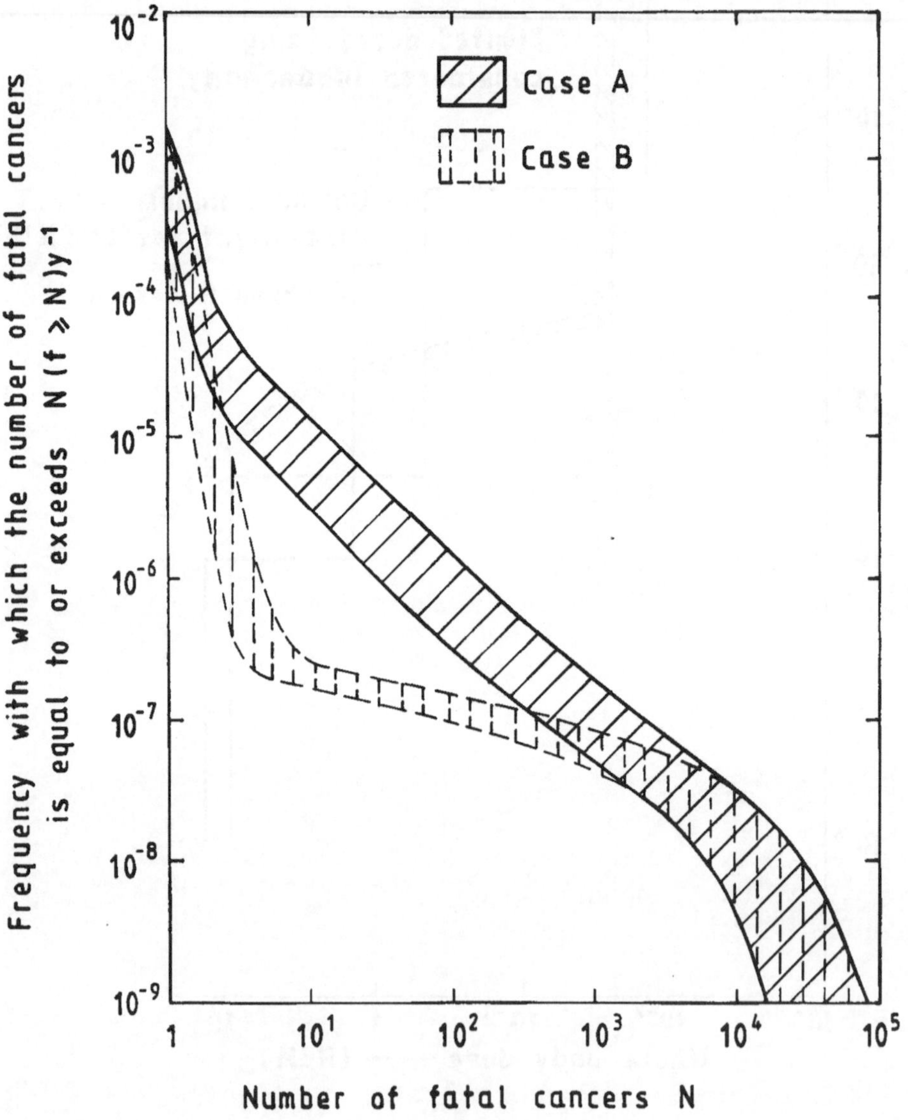

FIGURE 7.4
Severe Core Damage Frequencies from NUREG-1150
(Source: USNRC, 1987)

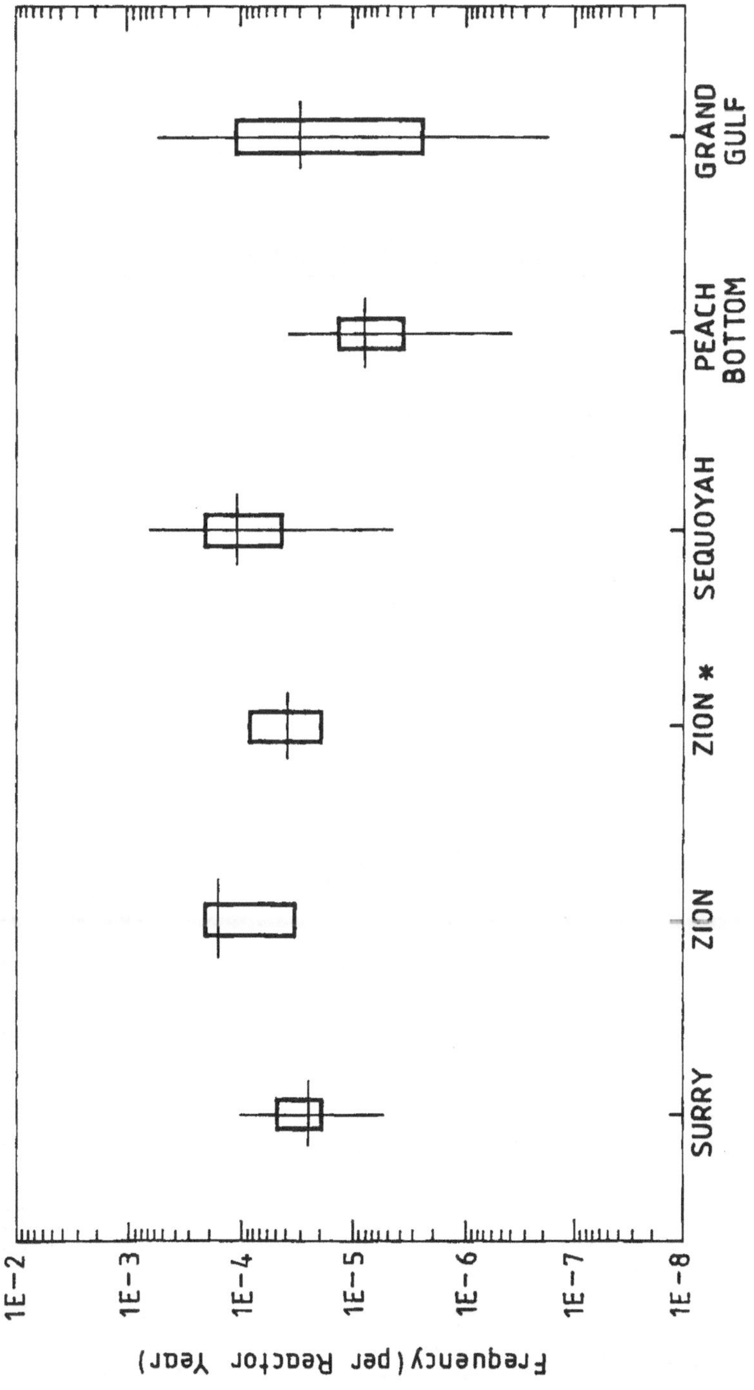

* ZION core damage frequeny with reduced CCW pipe failure rate, see section 3.2

FIGURE 7.5
Early Fatalities from NUREG-1150 and the Safety Goal
(Source: USNRC, 1987)

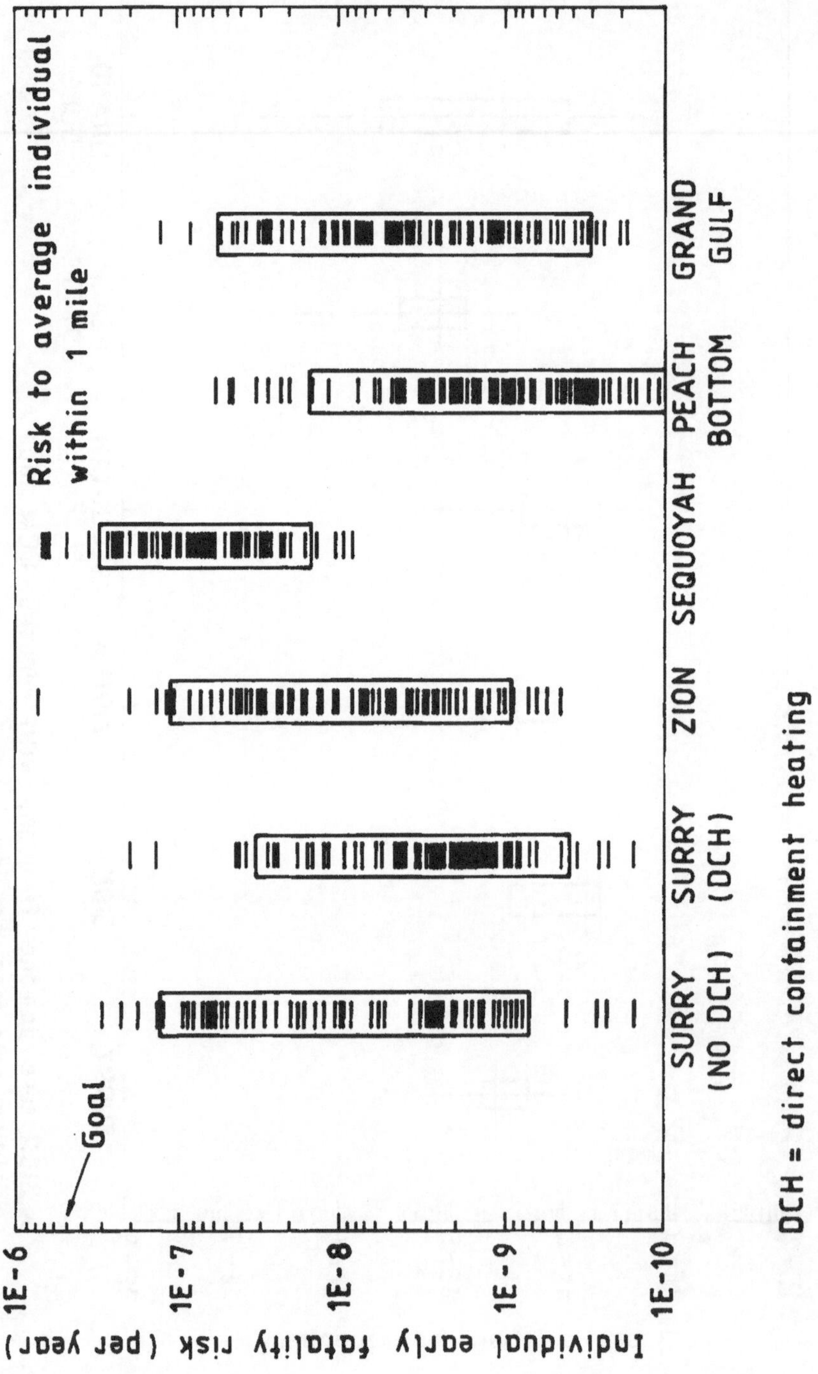

FIGURE 7.6
Cancer Fatalities from NUREG-1150 and the Safety Goal
(Source: USNRC, 1987)

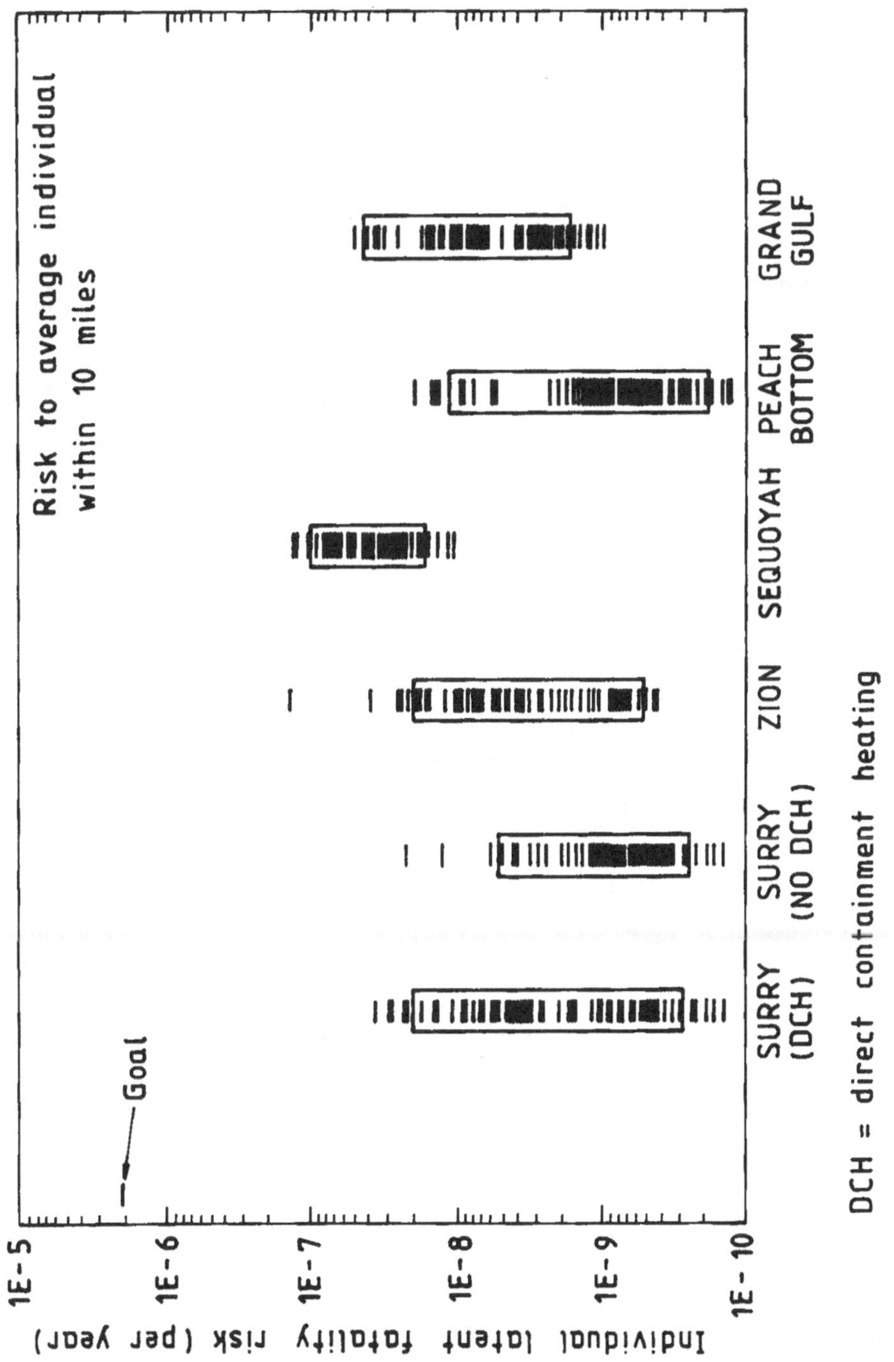

CHAPTER 8

MAKING RISK MANAGEMENT DECISIONS

8.1 Introduction

As we said in Chapter 1, risk management is concerned with taking decisions about risk tolerability. In this chapter we examine how this can be done. As a result we shall be concerned both with the decision process and the way in which quantitative targets can be set. We are able to make recommendations on both these issues.

The starting point is the theoretical basis of decision making which is described in subsection 8.2.1. Although the methods are not widely applied in practice for the reasons outlined in subsection 8.2.2, the ideas form a useful background to the other concepts discussed later. Sections 8.3 and 8.4 look at how decisions have been, and continue to be made in practice in the UK: section 8.3 considers how estimates of risk have been taken into account by the inspectors at several public inquiries, whereas section 8.4 is concerned with the form in which quantitative targets have been set in the nuclear industry. Our recommendations on appropriate forms for safety goals, and the way in which they should be implemented spring from this discussion.

8.2 Theory of Decision Making

8.2.1 Expected utility theory

Modern decision theory is unambiguous about how decisions under uncertainty ought to be made by the single decision maker. First the possible courses of action are listed, then the possible outcomes of each course of action. Next a numerical estimate of the worth of each outcome, called the utility, is assigned as is an estimate (objective or subjective as appropriate) of its probability. The decision maker then takes that course of action which maximises the expected utility. This approach is described by Lindley (1975), who makes a strongly stated case for its use in all decision making situations.

The two essential ingredients for this approach are thus the representation of uncertainty by means of probabilities and the measurement of dissimilar quantities (financial losses, health effects, benefits of a given technology.....) on a common numerical scale. This scale could be a financial one in which case the technique is referred to as cost benefit analysis (CBA). In practice the difference between CBA and the general decision theoretic approach is that CBA will often be restricted to pure economic issues, and involve, perhaps, concepts such as discounting, whereas it is axiomatic that all matters be included in the maximum expected utility approach, including political and perceptual considerations if these are felt to be relevant to the decision.

An example of the difference between CBA and the more general approach is the fact that the utility of money is a convex function for most people so that greater value is assigned to one's first million than the next. Such utility functions are termed 'risk averse' and this is consonant with the normal use of this term in the nuclear context where high consequence accidents are assigned a disproportionately larger value (disutility) than low consequence events. In this report we have preferred to use the term 'high consequence aversion' as a more exact description of the effect.

Another important aspect of applying these optimising techniques is the scope of the courses of action. This could be global, for example in deciding whether or not to build a power station, in which cases the benefits of the station would have to be compared with the disbenefit of the risk. Alternatively the optimisation could be local or marginal, for example in deciding whether to fit an additional safety system, in which case the extra cost is compared with the averted risk. There is little fundamental difference between the two cases, but the important practical difference is that the marginal analysis (in the present context) does not require quantification of the benefits of the hazardous activity, though in both cases a quantification of the risk in terms of a single number is needed. An approach to CBA for accidents is outlined in Appendix 1.

The legal description of 'reasonably practicable' given in Chapter 7 is immediately recognisable as being closely related to decision theoretic techniques, speaking as it does of balancing the quantum of risk against the sacrifice involved in averting it. For this reason it has been suggested by some authorities that ALARP/ALARA be implemented in terms of CBA in the context of operational exposure to radiation (ICRP, 1983; NRPB, 1986; UKAEA, 1987). However, it should be recognised that it is a marginal analysis which is involved (optimisation conditional on the activity being carried out) and that the legal definition of ALARP speaks of gross disproportion. This last is not thought to be the case for ALARA and it is this which is generally associated with CBA. Finally, it should be noted that both ICRP and NRPB emphasise that CBA is only an aid to judgement in making decisions, and thus forms only part of a complete ALARA approach.

It can be seen that these techniques address societal risk in that they use measures integrated over a population for decision making purposes. They can be used by society to manage individual risk only by disproportionately valuing higher individual risks in the way shown in Figure A1.1. We argued in Chapter 2 that the management of individual and social risks were done for different reasons, and it follows that careful consideration must be given when combining the two purposes in a single calculation. Nonetheless, the main point to be made is that the legal description of ALARP given in Chapter 7, and the CBA implementation of ALARA, are essentially societal risk management concepts.

Expected utility theory and CBA are by no means the only analytic decision aids available: others include cost effectiveness analysis, multiple criteria analysis, risk benefit analysis and environmental impact assessment. These do not represent fundamentally different approaches, but they do focus on different aspects and leave certain factors unquantified to various degrees. The proliferation of these methods is a consequence of the problems with the theoretical approach which are outlined in the next sub-section.

8.2.2 Problems

The techniques described in subsection 8.2.1 are rarely applied in practice for controlling technological risk. In view of the theoretical attractions of the method, which can be summarised in terms of 'coherent decision makers' (Lindley, 1975),it is important to understand why this is. In fact a number of reasons can be identified; they range from mathematical theorems through attitudes to risk to difficulties in quantification. It seems unlikely that all these reasons are completely independent, but any investigation of this would be beyond the scope of this report and we simply give brief descriptions of each.

The first, theoretical, point is the restriction of the theory to single decision makers. In fact there is a result, Arrow's Impossibility Theorem (see, for example, McLean (1987)), which says (in this context) that the risk preferences of society as a whole cannot be constructed in a sensible way from the preferences of its individual members. The conditions for this to hold are open to discussion, but this gives a scientific indication of the need for a political dimension to risk management. Thus efforts to quantify the attitude of various interested groups to decision options (see, for example, Edwards and von Winterfeldt, 1987) cannot provide the ultimate decision in cases of conflicting interest, though they provide valuable information for attempts to establish common ground and to reach compromise, if not consensus.

Decision theory provides a means to establish probabilities and utilities which consists of forcing subjects to express preferences between various wagers. The attitude of people to these wagers has been well researched (Machina, 1987) and it has been found that these attitudes are incoherent, that is, they do not align with the axioms of decision theory and can in principle result in a certainty of making a loss. Quantitative models have been constructed which better describe individual attitudes to risk and uncertainty, but they do not appear to be useful as prescriptions for societal decision making. In any case it is doubtful whether these attitudes are actually reflected in decision making since quantitative methods are rarely used.

This brings us to what is perhaps the most formidable problem in using quantitative methods: the aversion or difficulty associated with assigning the required numbers. This is best seen with respect to utilities: it can be thought amoral to assign a value to human life; it can be seen as difficult to quantify the benefit of an experimental programme; it can be considered nonsensical to treat a political factor in this way. The probabilities, too, cause difficulty. If risk estimates gave single probability figures, these could be directly used in the decision process. But we have indicated in Chapter 5 that such estimates should have an uncertainty analysis associated with them, and that because of problems in the technique judgement should be exercised in using the results. Both these factors militate against straightforward probability assignments for decision making purposes.

In spite of the problems outlined above, decision analysis in the form of CBA appears to have the potential to be a practical implementation of ALARP/ALARA. This is discussed further in Appendix 1. This chapter proceeds by looking for alternative procedures, using existing practice as a guide.

8.3 Public Inquiries in the UK

In this section we consider the experience gained from submission of the results of probabilistic safety assessments in evidence at public inquiries as reflected, in the reports of various inspectors. Particular attention is given to the impact of societal risk estimates and their use in formulating recommendations.

At the time of writing the final report on the EDRP inquiry was not available. The nuclear industry's experience of public inquiries, apart from the "Windscale" inquiry into the planning application for THORP (which predated much current thinking) is therefore limited to the Sizewell PWR inquiry. Some of the key points from the Layfield report have already been set out in Chapter 7 but further relevant conclusions are summarised below. However, there have been a number of public inquiries concerning non-nuclear "major hazard" plants which have taken evidence based on PSA and the experience at four such inquiries is compared with the Layfield findings. Two of these four inquiries concerned installations at Canvey Island (a proposed URL refinery and the British Gas Methane Terminal). The other inquiries concerned the construction of housing at Pheasant Wood, Thornton Cleveleys, Lancashire, in the vicinity of an ICI works handling chlorine and the construction of a natural gas liquids (NGL) pipeline from St Fergus to Moss Morran in Scotland.

8.3.1 The Sizewell inquiry

In this brief summary of key points references to the Layfield report are given in brackets.

Although CEGB paid little regard to societal risk in setting their design targets, Layfield concluded that both individual and societal risk should be considered in evaluating the risk from Sizewell B (2.101e). However, risk estimation was not a principal influence on his conclusions (2.18; that is, the results were sufficiently small that they were not crucial to the ultimate decision) and too much weight should not be placed on its results, which are subject to substantial uncertainty (2.61).

Layfield concluded that nuclear power was not unique as an industrial activity with the potential to kill thousands of people (2.124d). He considered that social risk criteria for nuclear power stations ought to be set (2.100) - and made a number of recommendations to that end - and that attempts to develop a common measure of social risk should be continued while bearing in mind its limitations (2.102f). In arriving at such criteria an equal weighting could be given to all deaths (12.23) or consequence aversion could be incorporated by making accidents involving large numbers of deaths proportionately more improbable (12.24). He referred to two criteria described in evidence; Kinchin's criteria which do not incorporate consequence aversion (36.77) while the Groningen criteria not only included a very marked consequence aversion but also considered that any risk of more than a thousand fatalities was unacceptable (36.73). Layfield found that he could not adopt the latter concept (36. 94).

In making his conclusions on the risks from Sizewell 'B' Layfield made his own assessment in which he weighted all deaths from beyond design basis accidents equally irrespective of the number killed by an accident and combined the results with those from smaller accidents and normal operation. He considered the result of up to one death every 80 years to be minute by any standards for an industrial enterprise of the magnitude and complexity of Sizewell 'B'. (This figure included risks to the public overseas: limited to the UK it was one death in 500 years). He also found that allowing for consequence aversion would not change this conclusion and that it would be difficult to justify spending more than an additional sum of about £0.5m to further reduce the risk to the public (47.53).

Layfield's conclusion that consequence aversion would not change his conclusions could be challenged. It appears to be dependent on some "realistic" risk aversion index, consistent with that of Okrent (1981), who considered that decisions on possible multiple fatality risks revealed that an aversion factor of much less than two and probably nearer to one reflected judgements taken in practice. In deriving his figures Layfield ignored early fatalities as they would be few compared with the number of fatal cancers. Unlike cancers early deaths could be directly ascribed to the accident causing them. However, it was argued in Chapter 6 (although it was not raised by Layfield) that late effects should not be subject to consequence aversion in the calculations even if early deaths are, as the delayed effects will not be clustered together in space and time leading to less societal impact.

In considering safety criteria, Layfield concluded that the opinions of the public should underlie the evaluation of risk and that Parliament was best placed to represent public attitudes (2.101h).

8.3.2 Interpretation of individual risk at other inquiries

Layfield found no authoritative guidance on levels of individual risk which could be considered negligible, but considered that a risk of death of one in a million years was likely to be broadly tolerable provided that there were associated benefits to justify it. Inspectors at three of the other inquiries considered here effectively concluded that there was no absolute standard or value of risk which could be considered negligible. However there were some important differences from Layfield's views, as described in the following paragraphs.

Inspector de Piro, reporting on the Canvey Methane Terminal, concluded that only orders of magnitude were important in taking decisions and that the risk from the terminal was low by any standards and small in comparison with those in everyday life: the peak individual risk was about 60×10^{-6} per year. He also found it unnecessary to consider benefits in reaching his conclusion about acceptability. Although he recognised that "theoretically or philosophically" imposed risks could not be justified without benefits he claimed that "in real terms" he had been able to consider the risk alone. Indeed, he stated that he would have found it difficult, or even impossible, to balance risks which were otherwise unacceptable against economic or strategic benefits. In contrasting this with Layfield it is worth noting that the levels of risk which de Piro found to be small were much higher than those from Sizewell 'B' and it is understandable that he might have found that significantly higher risks could not be justified even by significant benefits, as they might have exceeded the 10^{-4} per year figure given in the First Report of the Advisory Committee on Major Hazards as a possible upper limit for acceptability. Interestingly, de Piro rejected suggestions that "nuclear standards" should be applied to the methane terminal. In contrast to Inspector de Piro, Inspector Ward, at the previous Canvey Inquiry into a proposed refinery, considered that individual risks of the order of 10^{-5} per year could be balanced by the benefits, particularly since the risks were likely to have been overestimated.

The Inspector in the Pheasant Wood Inquiry also found it difficult to balance planning gains from housing construction against risks, stating that the gains might be bought too dearly even at risks of the order of one in a million years. On the other hand, he cautioned against an excess of prudence, which would waste resources. Nevertheless, he clearly considered that individual risks in the range $1\text{-}50 \times 10^{-6}$ per year were minimal and acceptable, being of the order of risk which was inevitable even from the best run chemical works.

The Moss Morran pipeline inquiry reporter clearly felt that individual risks of the order of one in a million years were of low order, but implied that best practicable means were necessary in handling NGL and even then a judgement was needed about whether what was practicable was good enough. In line with this view he suggested that the pipe wall should be thicker adjacent to vulnerable population notwithstanding the low level of risk.

8.3.3 Societal risk at other inquiries

Layfield considered his expectation value of one death in 80 years to be minimal by industrial standards, without giving any yardstick against which he had drawn this conclusion. Inspectors at the other inquiries considered here, while considering societal risk important, were unable to draw firm conclusions on societal risk estimates presented to them. Indeed, there has been a lack of clarity in the evidence on the purpose of presenting individual and societal risks which has left the inspectors with little guidance on which to make recommendations.

A major issue in the Canvey inquiries, in particular those concerning the proposed URL refinery, was whether the presence of a number of installations in close proximity, each of which might present acceptable risks in their own right, led to levels of risk which, in total, were unacceptable. This focusing of risk on specific populations is revealed in their levels of individual risk. (The total societal risk would be essentially the same if the plants were distributed around the country, given that interactions between the plants were not found to be of significance). This feature was not brought out clearly in the evidence and Inspector Ward appears to have been primarily influenced by societal risk in recommending that, although the extra refinery woud not of itself present an unacceptable risk, the background risks were already of concern and should not be increased. Ward's yardstick against which he compared the chance of accidents causing 10 or more casualties was the average individual risk of death in a motor accident. Finding the former to be an order of magnitude greater than the latter weighed heavily with the Inspector. He recognised that this compared individual and societal risks but clearly needed some context to interpret the societal risk estimate.

In the succeeding Canvey inquiry into the methane terminal, Inspector de Piro seemed to be mainly influenced by the order of the individual risk estimates. He did compare the chance of more than 10 casualties from methane terminal accidents with the risk of flooding, which he found to be of similar order. As the Inspector doubted whether the residents were now worried about flooding (the sea defences having been raised following major floods in 1953) he found no reason to be influenced by the societal risk from the methane terminal. It is notable that the Inspectors at both of these Canvey Island inquiries only considered the bounding information given to them on societal risk, that is, the chance of exceeding 10 casualties. No attention whatsoever was given to the low frequency-high consequence contributors to these figures.

In the Pheasant Wood inquiry the issue was whether to build more houses near to an ICI works. As no safety concerns had been raised the level of individual risk at the existing housing was implicitly accepted. The additional societal risk from construction of the houses was therefore the central issue. HSE, in objecting to the proposal, chose to test their pragmatic guidelines on separation distances, rather than using a full PSA in evidence. In this they were criticised by the technical assessor, who went on to make his own estimates of individual risk based on the evidence given by various parties. On this basis the inspector concluded that the risks were of low order (see above). The assessor did not indicate how he would have interpreted societal risks had estimates been given in evidence.

Similarly, it could be claimed that the Moss Morran pipeline inquiry should have addressed societal risks. The levels of individual risk were considered to be of low order, but this is not surprising for a transport type of operation, where exposure at a given point is limited. However, the frequency of failure somewhere along the route is usually high compared with a static installation and this can only be considered through the societal risk. HSE and Shell chose not to submit societal risk estimates and the inquiry did not seek to elicit them.

Although the final report of the EDRP Inquiry is not yet available, it is interesting to note that considerations of societal risk occupied a substantial place in the inquiry debate. The applicants' (UKAEA and BNFL) case was based on a design accident risk target of 10^{-7} per year for the most exposed members of the public (a suitable fraction of the site 10^{-6} per year target). This design approach to risk management was condemned by the principal objectors (the Orkney & Shetlands Joint Islands Council (JIC)) as failing to assure protection against societal risks: multiple delayed cancer fatalities and agricultural produce restrictions in particular.

The JIC arguments were countered by the applicants on the basis that

(i) full scope societal risk management was impractical at the early design stage, individual risk providing a control measure which was far simpler to apply;

(ii) comparison of social risks was not straightforward, involving value judgements - for example, as to the relative importance of different levels of consequences. There were no widely accepted reference levels of societal risk against which comparisons coud be made;

(iii) application of the design individual risk target would in any case afford a very high implicit degree of protection against societal risks, with some exceptions which were well understood and were inherently protected against in the EDRP design.

8.3.4 Public perceptions

Inspector de Piro (Methane terminal inquiry) considered that both actual or objective risk and perceived or subjective risk were relevant in a planning decision. He concluded that there was fear and anxiety amongst the local population but that this was not justified. He therefore concluded that steps should be taken to reassure the public that they did not live in a dangerous place and that comprehensive emergency planning should be instituted. (The inquiry predated legal requirements for such plans). He also concluded that it would be wrong for the terminal to be expanded in the future, although British Gas did not have any plans for expansion.

In the Pheasant Wood housing/chlorine plant inquiry the Inspector considered that informed local views were important in issues of societal risk. He had registered some difficulty over whether small risks could be balanced by planning gain (see above) and the only risk data presented was on individual risk. In this case the "informed" local views supported the plant and were not concerned about safety. Nevertheless, these views were presumably based on perceptions of the probability of an accident occurring rather than the differential societal risk which, as noted earlier, was at the crux of this inquiry.

8.3.5 Conclusions

It is important that the views of individual inquiry Inspectors, which may or may not have influenced decisions taken by government ministers, are not given too much weight. Nevertheless, on the basis of patchy and limited experience, the following conclusions might be drawn:

(i) The importance of societal and individual risks in specific decisions has not generally been clearly presented in evidence: the numbers have effectively been left to speak for themselves. As a result, inquiries have not always focused on the most relevant aspects in particular cases.

(ii) Nevertheless, Inspectors have considered societal risk to be an important factor. However, they have struggled to interpret it in the absence of any yardstick for comparison. Individual risks have proved easier to interpret by conventional comparisons with other risks, and have therefore often been of greater influence.

(iii) Although Inspectors have considered societal risk to be important, they have not shown any indication of aversion to high consequence accidents.

(iv) There has been a degree of consensus that public opinion, and therefore risk perception, is an important factor in planning decisions.

8.4 Risk Targets

In Chapter 6 we recognised that decisions about tolerable levels of risk could be made in two ways. Either they should be outweighed by the associated benefits, or they should be below some threshold for concern. Theoretical means of decision making, and quantitative decision aids, focus on the first of these - the comparison of risks and benefits, as outlined in Section 8.2. The 'unique benefits' and 'equivalent benefits' approaches described in Chapter 6 simply restrict the range of options to be considered by such analyses. But the discussion of existing targets in Chapter 7 and the attitude of inquiry inspectors outlined above shows that in practice the second approach - risks of no concern - is generally favoured. In this section we discuss the form of possible quantitative safety goals in the light of this. Thus a general format of safety goal (which we call a banded target) is described in subsection 8.4.1 and this is used to interpret the way other targets have been set. These include most of those described in Chapter 7 and add the CEGB's Design Safety Criteria (CEGB, 1982a) which we have not so far described in any detail. The problems which arise in assessing compliance with such targets are then discussed in subsection 8.4.2.

8.4.1 Banded targets

The discussion in Chapters 6 and 7 shows that there are three basic ingredients of quantitative safety targets. A given approach would not necessarily include them all, and the precise interpretation and implementation of each could vary.

- Levels of risk which are intolerable. Such levels could refer to individual risk in terms of the various types of personal risk listed in Chapter 3, and they could also refer to social risk, in terms both of collective personal risk and societal financial risk. The bounding levels could be single numbers, or, in the case of social risk, f-N lines, perhaps incorporating risk aversion concepts. The grounds for intolerability could be legal, an assertion of "de manifest is" (Travis et al, 1987), or be based on the risk comparison principles set out in Chapter 6.

- Levels of risk which are small in some sense. This sense could be one either of triviality (or de minimis), of being of no concern to the average person (Ashby, 1978), of acceptability (ICRP, 1977), of being a design target, or of being likely to be ALARP/ALARA. Again the risk could be individual and/or social.

- A requirement that risks should be ALARP or ALARA. This could be purely judgemental or based on a quantitative approach such as CBA. An important difference is that here (in the quantitative case) only a single calculation is carried out, and the various types of risk are not considered separately. The legal definition of ALARP is essentially local optimisation (that is, it is assumed that the project will go ahead), but a global optimisation (checking that the benefits outweigh the risk) would also be possible in principle. From now on we shall not distinguish between ALARP and ALARA and use only the former in place of either.

A fourth concept which has not been included above is that of as low as technically achievable (ALATA). There is no *official* precedent for such an approach and Layfield, for example, cautioned against disproportionate expenditure on safety. Indeed, there is probably no realistic means of implementing such a concept (short of cancelling the project) since it would result in unlimited expenditure.

In general it can be expected that ALARP is applied between the intolerable and 'small' levels, depending on the definition of 'small'. One exception to this is noted below. This introduces the concept of a 'banded' safety goal; see, for example, Figure 3.10 which which shows the Dutch criteria in terms of a 'reduction desired' band separating the 'acceptable' and 'unacceptable' regions.

The approaches to risk management which we have examined in this and previous chapters can be characterised by the mix of the above ingredients which they contain. This is now discussed in more detail for a number of approaches by way of clarification.

<u>NII SAPs</u> The fundamental requirement is ALARP, implemented judgementally. There are also the assessment reference levels - small levels of frequency and dose for single accidents - which can be interpreted as a CCDF for individual risk and integrated to give total individual risk (Harbison and Kelly, 1985). However, these levels are not small in the sense that the risk is judged likely to be ALARP, but in the sense that the assessment of the safety case is likely to have been carried through as far as practicable (see previous discussion in sub-section 7.4.1). This picture is thus not quite one in which ALARP is applied above a lower limit, though in practical terms the above distinction will have little effect.

<u>CEGB's Design Safety Criteria</u> The CEGB's Criteria (CEGB, 1982a), like the SAPs, are concerned only to a small degree with quantitative targets. Again like the SAPs they set maximum allowable frequencies for three dose bands below 1 ERL, though the frequencies refer to the sum of all the accidents giving doses in the band, not single accident sequences. Furthermore the way in which the dose is calculated varies between the CEGB and NII (Lange et al, 1987). Unlike the NII, the CEGB has a frequency target for "uncontrolled releases" of 10^{-6} per year. This includes all releases over 10 ERL, though not necessarily those over 1 ERL. The targets are derived from a level of individual risk that is "lower than the everyday risks of life that currently exist, and can be compared favourably with other risks for similar types of activity that society finds reasonable". Further numerical targets can be found in the CEGB's Design Safety Guidelines which are derived from the Criteria for each reactor type (see, for example, CEGB, 1982b). These relate to the reliability of specific systems. All these levels are targets and it is recognised that "it is not essential that the guidelines be met in all respects in order to ensure adequate safety". Thus the CEGB's policy is one of defined lower levels which produce individual risks of 'no concern'.

<u>Layfield</u> As stated above, Layfield's recommendation appears to have been reached in terms of accepting that the individual risk lay below an intolerable level, and requiring that the benefits outweigh the risks, a global optimisation.

HSE Tolerability of Risk Document It is clear from the discussion in subsection 7.4.3 that the HSE (1988) have a full banded target in mind for individual risk to members of the public, whereas for worker risk and societal risk, upper limits only are derived to stand alongside the ALARP requirement.

Farmer Criterion (see Chapters 3 and 7) Although this addresses societal risk to some extent, it does not fit in well with the picture described since it relates to individual accident sequences (though it has been interpreted as a release CCDF in a way similar to the SAPs (Beattie et al, 1969)) and little guidance is given as to the interpretation of the levels. The discussion in Farmer (1967) suggests they can be considered to be somewhere between small and intolerable; ideas of banded goals are relatively recent.

UKAEA Code of Practice and Safety Directives (see Chapter 7 and Appendix 1) As applied to routine exposure, the Code of Practice works in terms of small levels of individual risk, with a cost benefit implementation of ALARP above this along the lines suggested by the ICRP and NRPB. Finally, such exposure is subject to legal limits, so all three ingredients are seen to be applied here to individual risk.

USNRC Safety Goals (see Chapter 7) These goals refer to levels of average individual risk of early death within 1 mile and average individual fatal cancer risk within 10 miles. Their original formulation differed from this in some respects, but the most important difference was the inclusion of a cost benefit guideline (USNRC, 1983) in cases where plants did not achieve these levels which were termed design objectives. The version which has recently been implemented (USNRC, 1986) has dropped both the cost benefit guideline, and the term design objectives; however, it has introduced a 'general performance guideline' relating to the frequency of large releases. The original version would thus appear to consist of small levels and a quantitative implementation of ALARP (Higson, 1985). Guidance to USNRC staff on how to apply the new safety goals has yet to appear, but they seem to be much closer to a situation where the small levels and the intolerable levels coincide, with little intervening scope for ALARP.

Dutch Provisional Criteria (see Chapter 7) These are full banded quantitative safety goals set in terms of individual risk and societal risk of early death represented by CCDFs (Versteeg and Visser, 1987). Risks are unacceptable above the upper levels and acceptable below the lower levels. Risk reductions are desirable in the intermediate region, though quantitative criteria for this are not set out.

These and other examples show that almost all extant quantitative statements of risk management policy can be framed in the way suggested at the start of this section. The number of ingredients used, and the types of risk to which they are applied, vary. In particular there is very little which addresses societal risk.

8.4.2 Assessing compliance

The introduction of large and small levels of risk into quantitative safety goals leads to a new theoretical problem. This relates to how to decide on compliance with such targets when estimated risks are subject to uncertainties, both explicitly represented and implicit, as set out in Chapter 5. This problem is generally tackled by the specification of a set of rules as to how compliance is to be assessed. In general the rules will differ depending on whether it is thought possible to give a reasonable quantitative representation of the uncertainty. For example, it is intended that the levels in the USNRC Safety Goals should be compared with mean values (USNRC, 1986), which presupposes a quantified representation of uncertainty. Alternatively the concept of 'probability of compliance' is introduced by Hayns and Unwin (1985) as one of many mathematical options which are available for explicitly stated uncertainty. By contrast the UK tradition is one of using reasonable pessimism to get single values to be compared with targets; this is founded on a belief that uncertainties cannot be sensibly estimated or that the implicit uncertainty dominates. Of course criteria are not set in a vacuum and result from a careful consideration, iterating between operator and regulator, of what is reasonable in the light of what can be calculated by the PSA method on the one hand, and what might be tolerable levels of risk on the other.

To summarise, compliance with fixed targets can only be addressed in the context of the chain of reasoning which has led to the targets and the methods, approximations and judgements used to estimate the risk. Mathematical prescriptions cannot form the basis of such decisions, though they may, perhaps, provide useful supplementary information.

8.5 Conclusions and Recommendations

(1) Formal decision theoretic techniques for balancing risks and benefits are beset with difficulties, though they do offer the advantage of clarifying the reasoning behind decisions.

(2) Public inquiry inspectors have considered societal risk to be important, but have struggled to interpret estimates usefully. They have shown no sign of 'high consequence aversion'.

(3) So-called banded targets consist of levels of risk which are high in some sense, levels which are low in some sense, and a requirement for the reduction of risk so far as practicable in the intervening region.

(4) Many existing quantitative risk management policies can be interpreted as complete or partial banded targets.

(5) The use of banded targets is recommended so far as possible.

(6) Formal decision theoretic techniques for balancing risks and benefits should not be mandatory.

(7) However, if risk managers wish to use such a technique, a method to estimate the cost of risks to the health of the public is described in Appendix 1.

(8) In using risk estimates to assess compliance with quantitative safety targets due account should be taken of the explicitly represented and implicit uncertainties associated with the estimates. It is not recommended that any particular mathematical technique to do this be mandatory.

8.6 References

Beattie, J R, Bell, G D, and Edwards, J E, 1969, **Methods for the Evaluation of Risk.** UKAEA Health and Safety Branch, AHSB(S) R159.

CEGB, 1982a, **Design Safety Criteria for CEGB Nuclear Power Stations.** HS/R 167/81 Revised.

CEGB, 1982b, **PWR Design Safety Guidelines.** CEGB Generation, Development and Construction Division (DSG-2, Issue A).

Edwards, W, and von Winterfeldt, D, **Public Values in Risk Debates.** Risk Analysis, 7, pp 141-158.

Farmer, F R, 1967, **Siting Criteria - A New Approach.** IAEA Symposium on the Containment and Siting of Nuclear Power Reactors, Vienna 3-7 April, 1967. IAEA SM-89/34.

Hayns, M R, and Unwin, S D, 1985, **Rational Quantitative Safety Goals.** ANS Winter Meeting, San Francisco.

HMNII, 1979, **Safety Assessment Principles for Nuclear Power Reactors.** HSE, HMSO, London.

Higson, D J, 1985, **Nuclear Reactor Safety Goals and Assessment Principles.** Nuclear Safety, 26, pp 1-13.

HSE, 1988, **The Tolerability of Risk from Nuclear Power Stations.** HMSO, London.

ICRP, 1977, **Recommendations of the Commission.** ICRP Publication 26. Ann ICRP, 1, No 3.

ICRP, 1983, **Cost Benefit Analysis in the Optimisation of Radiation Protection.** ICRP Publication 37, Ann ICRP, 10, No 2/3.

Lange, F, et al, 1987, **Safety Objectives for NPPs in Terms of Dose-Frequency Targets.** PSA'87, International Topical Conference on Probabilistic Safety Assessment and Risk Management, 30 August - 4 September 1987, Zurich, TUV Verlag.

Layfield, Sir Frank, 1987, **Sizewell 'B' Public Inquiry Report.** HMSO, London.

Lindley, D V, 1975, **Making Decisions** (2nd Edition). Wiley.

Machina, M J, 1987, **Decision Making in the Presence of Risk.** Science, 236, pp 537-543.

McLean, I, 1987, **Public Choice.** Basil Blackwell, Oxford.

NRPB, 1986, **Cost Benefit Analysis in the Optimisation of Radiological Protection.** ASP 9, HMSO, London.

Okrent, D, 1981, **Industrial Risks.** Proc Roy Soc, A 376, pp 133-149.

Travis, C C, Richter, S A, Crouch, E A C, Wilson, R, and Klema, E D, 1987, **Risk and Regulation.** Chemtech, August 1987, pp 478-483.

UKAEA, 1987, **Code of Practice and Guidance Note: Radiological Guidelines for the Design and Operation of UKAEA Plant.** Safety and Reliability Directorate Report, SRD R 456.

USNRC, 1983, **Policy Statement on Safety Goals for the Operation of Nuclear Power Plants.** Federal Register, 48(50): 10772-10781.

USNRC, 1986, **NRC Adopts Policy Statement on Safety Goals.** Nuclear Safety, 27, No 4, pp 555-556.

Versteeg, M F, and Visser, B J, 1987, **A PRA Guide for the Netherlands; a Consequence of the Dutch Policy on the Risk Management Applied to Nuclear Energy.** PSA'87, International Topical Conference on Probabilistic Safety Assessment and Risk Management, 30 August - 4 September 1987, Zurich, TUV Verlag.

CHAPTER 9

THE OVERALL APPROACH TO SOCIAL RISK MANAGEMENT

9.1 Introduction

The purpose of this chapter is to draw together the lessons gained from the preceding chapters to see what guidance can be given on the principles of formulating and implementing risk management policy with particular reference to the social impact of major accidents. To this extent it is a summary of what has gone before. It is not the intention to extend this guidance to the point of providing specific recommendations for given activities.

We begin by discussing safety goals, particularly quantitative ones, in section 9.2. It is clear that such goals need to specify not only numerical levels to be achieved, but also how the risk is to be estimated and how the comparison is to be made. These and other practical issues are discussed in section 9.3 which also speculates on how a risk management policy can be evolved from existing practice. Finally section 9.4 briefly covers some of the organisational aspects which arise from this. The conclusions and recommendations are summarised in section 9.5.

9.2 Safety Goals

Technological risks are imposed on society because the benefits which result from the technology outweigh the risks. This is the fundamental social safety goal. As we saw in Chapter 8, this concept forms the basis for formal decision making techniques. At this basic level the situation is well defined.

However, the discussion in Chapter 8 also showed that these techniques are very difficult to apply in practice. The reasons for this are wide-ranging, but there are two main ones. The first is the difficulty experienced by individuals in assigning numerical values to the possible outcomes such as many deaths and forced evacuations. The second is the impossibility of combining such individual weighings to provide a course of action agreeable to society as a whole.

In these circumstances the nuclear industry has attempted to take a different approach to risk management. Rather than justify a level of risk as tolerable from an activity because the benefits are sufficiently large, it has tried to define levels of risk which are of no concern. It can then be argued that the imposition of such risks does not require detailed evaluation of the associated benefits, and the problems are avoided. The concept of levels of risk which are of no concern derives from the more basic 'tenets' expounded by Gittus and Hayns (1987) that: "if it can happen then it must not matter" (consequences of no concern) and "if it matters then it must not happen" (frequencies of no concern). At the same time it may also prove possible to identify levels of risk which are intolerable whatever the benefits, and these can also give useful guidance to risk managers.

The principles which can be applied to derive these large and small levels are described in Chapter 6. They show that there are many problems associated with this approach too. Some members of society will argue that no risk is tolerable under any circumstances, and others will see no benefit in certain activities (such as producing energy by nuclear fission) and again call for zero risk (perhaps not recognising that the risks of no electricity production are very large). The result of this is that risk tolerability is an essentially political matter, and risk managers can rely only partially on technical solutions to their dilemmas.

However, this report is concerned with these technical aspects and this means that we must focus on the three ingredients identified above:
-technical means for balancing risks and benefits,
-levels of risk of no concern, and
-intolerable levels of risk.
Together these ingredients constitute 'banded' safety goals.

The first of these has been discussed in Chapter 8 and specific recommendations are made in Appendix 1 as to how to apply cost benefit analysis in the accident situation. The discussion there is limited to health effects, and Chapter 3 makes clear the need to consider other types of risk including forced evacuation and the financial consequences of land contamination, food interdiction, resettlement and so on. These can in principle be added in to the calculation. An important point to be aware of is that only a single calculation is carried out: all relevant factors have to be included.

This is not the case for large and small levels of risk which can, in principle, be derived for the different types of risk, and, indeed, for different representations of the same type of risk. For example, for early death we try to control the individual risk, the integrated social risk and the frequency of accidents in each consequence range by using a CCDF target. Many quantitative targets appear to be rather arbitrary in nature and in Chapter 6 we set down a number of principles which can be applied so that such levels of risk can be derived on the basis of clear reasoning. Chapter 6 considers only reference levels of risk; the step from these to levels of no concern or intolerable levels is a matter of judgement on which little technical guidance can be given.

Some progress is made in Chapter 6 towards defining reference risk levels for early and delayed death. Furthermore, injury and birth defect risks from nuclear accidents can be managed as part of a strategy for managing the risks of death. However, reference levels for personal harm such as forced resettlement, and financial losses to the individual and the community are not available.

As for 'high consequence aversion', there is no reason at all to introduce this concept for managing the societal risk of cancer from accidents. For early deaths the situation is less clear; there is no objection to this concept in principle, but there is no theoretical basis, nor any information from risk comparisons, which enable reference CCDFs to be derived. This last is sometimes taken as the only measure of social risk, but aggregated measures such as total expected numbers of deaths assume a greater importance in these circumstances. If there are felt to be political or social demands for CCDF targets, then more arbitrary reasoning can be applied to divide the overall risk between various consequence levels. However, in the case of nuclear accidents it seems unlikely that 'high consequence aversion' applied to early deaths could ever be a limiting factor for practical risk management.

More arbitrary targets may also be considered on various grounds not directly related to public risk; for example, because they are believed to represent good engineering practice, to avoid social nuisance or to protect the future of an activity (see Chapter 3). The reasoning behind such more arbitrary targets should always be made clear.

The elements of risk tolerability which can be regarded as soundly based are therefore few, and a cautious approach is called for. This could consist of attempting to meet such low levels of risk as it is possible to define (possibly supplemented by more arbitrary levels), and, at the same time, making some assessment that the risks are outweighed by the benefit. Formal methods, such as CBA, may be able to make some contribution here, but their use should not be mandatory, and a substantial element of judgement may still be required.

9.3 Practical Aspects of Nuclear Risk Management

The discussion in the previous section, and the line of this report in general, has been one of thinking about risk itself, divorced from the practicalities of: whether techniques are available to estimate the required quantities; if so, how compliance should be judged; and, given this, whether it is technically possible to design compliant plants. It also ignores existing practice; nuclear risk targets have been set in many ways, but, as Table 7.3 shows, rarely in terms of risk. This concentration on risk has been a deliberate policy because, in the final analysis, it is the events of basic concern - deaths, financial losses, disruption of society - which must be controlled. Targets related to other quantities need interpretation which can be confusing and variable. For example, radiation dose targets are irrelevant for non-nuclear hazards and imply different levels of societal risk for different sites. Within the nuclear area, accident release targets are irrelevant to managing the risks from waste disposal.

Nonetheless, future practice must evolve from what happens now, moderated by the technical capability of assessment methods, and in this section we briefly discuss some of these practical matters in more detail. Subsection 9.3.1 gives an overview of the existing situation, while in subsection 9.3.2 we consider how this might evolve towards a more risk-based approach.

9.3.1 Existing management of risk

The many targets shown on Table 7.3 are not actually as disparate as they seem. Much work has been done on interpreting these goals in terms of risk, particularly individual risk, and this has shown a considerable degree of uniformity. Of course this only applies in the situation for which the goals are intended, in this case nuclear reactor safety. In any case this is not uniformly true. For example, the Dutch design basis targets apparently imply such a low level of individual risk that it is difficult to extend them beyond the design basis using the individual risk targets as a guide. More importantly, effects which do not have to be evaluated against targets - external hazards in the Netherlands, human factors in the UK - may be important contributors to risk. Nonetheless an international consensus on a tolerable level of individual risk of around 10^{-6} per year is emerging, partly as a result of these interpretations.

Against this there is an effect which tends to show agreement where none actually exists. Figure 7.2 is an example. This shows apparent consistency between the dose frequency targets for the UK, France and Germany. However, such agreement may be illusory due to differences in implementation relating to: extension to higher doses, physical modelling of dispersion, assumptions about weather, assumptions about the habits of individuals receiving the dose and so on (Lange et al, 1987).

On the general question of what assumptions to make when estimating risk, we are quite clear that this should be done on a realistic basis. Many individual risk targets have been formulated in terms of hypothetical individuals, usually with extreme characteristics. These can only be regarded as approximations, and the underlying requirement is for the risk of the most exposed real individual to lie below the target. It is then possible to introduce approximations to ease calculation in specific cases. The same is true for societal risk estimates which should take account of likely long and short term countermeasures both on and off site. These would include the repair of components, operator recovery, evacuation, decontamination and interdiction of land and foodstuffs. It should be remembered that many of these, while reducing the risks to health, will give rise to significant financial risk. This risk may well exceed that of the averted health effects evaluated on a cost benefit basis such as that described in Appendix 1. This requirement may need significant development of risk estimation techniques.

For targets which are not set in terms of risk, there is no particular reason for realistic methods to be prescribed, since one in any case is not dealing with the effects of primary concern. Thus rules can be used to achieve some particular effect, or to reduce the importance of assessment uncertainty.

Reverting to the main discussion of the relationship between existing practice and risk, we find that the situation is even less well defined for societal risk. This report has been written because there is no clear view (except in the Netherlands) as to what tolerable levels of societal risk might be. Again much work has been done on the interpretation of existing targets, combined with site classification rules and emergency response plans. But, partly due to the great uncertainties in the assessment process, no obvious conclusions can be drawn.

9.3.2 Evolutionary approaches

There are two major reasons why targets tend not to be set in terms of risk. The first is that risk is a difficult concept to use, as we have seen, whereas release targets, for example, can be derived partly from considerations of technical practicability. The second is that low level targets are more easily applied by the designer and operator, using easier techniques with less uncertainty.

Thus an evolutionary approach might aim to set targets which are fairly close to existing practice, using quantities which can be reasonably easily estimated with a moderate degree of uncertainty. The problems of uncertainty could be further reduced if rules are made by the regulatory authority on how the remaining uncertainties should be dealt with. Against this it would be desirable to be able to develop a clear idea of the levels of individual and societal risk which might be implied, and for these to remain roughly constant over as broad a range of applications as possible.

One such evolutionary approach can be inferred from the recent HSE discussion document (HSE, 1988). As described in subsection 7.4.3, this seeks to control societal risk in terms of the frequency of accidents giving a specified dose at a specified distance. This is convenient to assess against, but does not lead to a clear idea of what levels of societal impacts might be implied; indeed these levels would vary from site to site, and depend on the accident spectrum of the particular type of plant. This situation could be improved by using a range of reference accidents of different severities, defined in terms of site-specific doses more closely related to actual societal impacts.

Another approach which has been suggested is to use a CCDF of individual dose as the primary target. Again this is closely related to existing practice, and is reasonably easy to assess against. It is left open as to whether pessimistic or realistic rules are used for this purpose and this would determine the extent to which individual risk would be explicitly controlled; a realistic technique would be directly related to individual risk, including that of early death. The degree of control of societal risk is less clear. Although it might be hoped that the variation of individual risk with distance from the plant would be relatively insensitive to type of plant, this approach would not take site dependencies into account. Thus it might be necessary to combine the individual dose CCDFs with site classification rules to manage societal risk, though this may not work well with cancer deaths which will mainly be induced outside the area to which these rules relate.

This immediately introduces another topic of considerable practical interest for societal risk management: the controversial area of de minimis doses (as opposed to de minimis risk as discussed previously). Because there is no direct evidence that exposure to radiation at levels well below background is harmful, it can be asked whether such exposures should be taken account of. On the one hand it seems prudent to do so, on the other it seems inappropriate to base risk management policy on hypothetical, intrinsically unobservable consequences. Of course this situation would change somewhat if a validated mechanistic model of radiation induced cancer were developed. The use of dose-based targets as described above focuses on the level of dose received by individuals and in any case it is desirable that the relevant information be made available so that risk managers and interested parties can assess the effect of a particular assumption. In this case the information would be a plot showing the number of individuals exposed to each dose level as shown in Figure 3.12. Returning to the main question however, we recommend on balance that risk managers should consider the use of a de minimis dose level so that their policy focuses better on the true social impact of accidents. This might additionally have the advantage of rationalising the spatial cut-off of consequence calculations which is often arbitrary.

To conclude, we do not wish to recommend any particular evolutionary approach to societal risk management. Our main concern is risk itself, and its control should always be the underlying motivation for risk management policies, though we recognise that these must be developed from present practice, taking practical constraints into account.

9.4 Risk Management and Society

It is clear that risk management problems cannot be resolved by scientific and technical methods alone. This means that a considerable burden falls onto society as a whole which has to seek its own solution using different techniques. It is important for the technical community to be aware of this and of how it can best contribute to the process. These issues were discussed at the symposium at the University of East Anglia mentioned in Chapter 2 (Roberts, 1988) and we briefly describe the main points here. They fall under two broad headings: the institutional arrangements for providing public safety; and the presentation of risk information.

The safety of commercial nuclear power plants in England is a complex matter, which does, indeed, confuse the public. The responsibility for their safe design and operation belongs to Nuclear Electric (NE). They must, however, present a safety case to the HSE (in the form of the NII) which is responsible for licensing plants and for ensuring that they are operated within the licence. NE applications to build a reactor may also be subject to a Public Inquiry. Such an Inquiry may consider safety issues in addition to planning and economic matters. The Inspector at a Public Inquiry can only make a recommendation: the final decision is made by the Secretary of State for Energy. He, and the Secretaries of State for the Environment and Employment (the minister responsible for the HSE), are accountable to Parliament and hence the public.

It is important that the public should have confidence in these arrangements, and the significance of this for the technical bodies involved is that their expertise and credibility should be accepted so far as possible. Unfortunately this is not necessarily the case at present. The reasons for this, and the ways in which this situation can be remedied, are beyond the scope of this report, but we recommend that further efforts be made to achieve this. The French example, quoted in subsections 7.2.3 and 7.5.1 appears to be a move in this direction.

One aspect of this is the desirability of making the dialogue between the plant operators and the safety/licensing authorities more open and accessible to the public. There are many things which can contribute to this. In the technical area they concern mainly the presentation of information generated during the estimation of risk. This includes not only the final estimates of risk in some form, but also the associated explicit and implicit uncertainties.

PSAs are large, complex and hard to understand. But the most important information they contain is normally fairly simple: the dominant initiating event; the system reliability or physical process to which some measure of risk is most sensitive; the important judgements which have been made to produce a result; and so on. This information should be made more easily available - to the safety analyst, as well as the risk manager and the public. The underlying details should also be as open as possible to the scrutiny of those who care to examine them.

Finally, the technical community should try to explain its need for quantitative targets as an engineering necessity. These targets may not be wholly acceptable to the public, so the underlying reasoning should be explained as carefully as possible. And if rather arbitrary assumptions are necessary on technical grounds or simply to account for some aspect of public concern, then this too should be explained.

9.5 Conclusions and Recommendations

(1) The fundamental social safety goal is that risks should be imposed only if the resulting benefits outweigh them.

(2) This is difficult to apply quantitatively. An alternative policy is to try to make the risks low enough to be of 'no concern,' and in any case to make sure they are not intolerable regardless of the benefit.

(3) In this case, risk management should in the first instance be based on levels of risk which can be shown to be low or high using reasoned arguments starting from reference levels of risk derived using the principles outlined in Chapter 6.

(4) Additional, more arbitrary, targets may be introduced on grounds which should be fully explained.

(5) Because we recognise that risk management policies must be developed in an evolutionary way from current practice, the primary targets may not always be phrased in terms of risk. Nonetheless the implied level of risk should always be considered.

(6) Risk estimates should be made on as realistic a basis as possible, taking into account the characteristics of the population at a specific site and the post-accident countermeasures which are likely to be taken. Estimates of financial harm should be undertaken appropriately.

(7) We recommend that risk managers should consider the use of a de minimis dose level and, in any case, dose bands should be used to present the results of radiological risk estimates.

(8) It is desirable for the institutional provisions for public safety to be as clear to the public as possible, and for there to be greater public accessibility to safety cases and the dialogue between the organisations involved.

(9) Part of this need can be met by improving the presentation of: the guidance required by technicians, engineers and operators; the risk estimation process; and the reasoning behind safety goals.

9.6 References

Gittus, J H, and Hayns, M R, 1987, **Risk Assessment.** Proc Roy Soc Edinburgh, 92B, pp 139-154.

HSE, 1988, **The Tolerability of Risks from Nuclear Power Stations.** HMSO, London.

Lange, F, et al, 1987, **Safety Objectives for NPPs in Terms of Dose-Frequency Targets.** PSA '87, International topical Conference on Probabilistic Safety Assessment and Risk Management, 30 August - 4 September 1987, Zurich, TUV Verlag.

Roberts, L E J, (Ed), 1988, **Risk Perception and Safety Targets for Major Accidents.** Report of a seminar held at the University of East Anglia, 16 October 1987.Research Report No 4, Environmental Risk Assessment Unit, University of East Anglia, Norwich.

CHAPTER 10

CONCLUSIONS AND RECOMMENDATIONS

This chapter brings together the conclusions and recommendations from all the other chapters under the headings of preliminaries (section 10.1), risk estimation (10.2), risk evaluation (10.3), risk management (10.4), and the overall approach (10.5). Minor changes have been made to improve readability and relevance. Recommendations are printed in boldface.

10.1 Preliminaries

In Chapter 2 we provided formal definitions of risk in terms of probabilities, but the general concept of risk is complex with historical, cultural and economic aspects. We found that the regulation of individual risk does not necessarily ensure tolerable levels of social risk in all circumstances.

Therefore social risk also needs to be controlled.

We recognised the needs of various parties - regulators, politicians and the public - but we stated that our main concern in this report will be with the problem of providing technically based guidelines to the engineers and scientists who actually build and operate plant. Thus we shall try to establish guidance which, when implemented, would result in broader risk acceptance criteria being met.

In Chapter 3 we examined various types of risk and the way they could be represented. Various recommendations spring from this:

Societal risk management should take into account at least five types of personal risk - early death, late death from cancer, serious injury, permanent or long-term evacuation and serious birth defects - as well as non-personal harm to society, which is principally financial.

It is not appropriate for primary quantitative risk targets to be set in terms of a single simple release parameter.

The CCDF (or F-N line) is recommended for evaluating social risk because it is the most accepted representation and the easiest to use and understand. This does not necessarily mean that targets should be set using CCDFs. The integrated measure of social risk can give a useful coarse description, although it sometimes conceals important information.

10.2 Risk Estimation

Chapter 4 estimates existing individual and societal risk levels in Great Britain and elsewhere. For individual risk we focus particularly on cancer and accidents (for comparability with nuclear risks), and occupational fatal and non-fatal injury. It is important to realise that the data used in the risk estimates are time dependent and the current trend is towards levels of increasing safety. Also the individual risk calculation is very sensitive to the choice of the population at risk.

The data for societal risk show that for many natural disasters most deaths take place in the highest consequence events observed whereas in man-made accidents they are evenly distributed through the consequence range, or are concentrated in low consequence events.

Chapter 5 examines how risks of major hazard plant are estimated probabilistically in the absence of accident data. However, we recognise that probabilistic methods, associated with quantitative safety goals, cannot provide the complete safety case for hazardous installations.

In order to incorporate probabilistic results into a risk evaluation, account must be taken of all the subjective factors which enter. As well as giving guidance on how the analysis is to be carried out, the risk evaluator must form a view on the extent to which he will accept these judgements, or what allowance to make for them in reaching decisions. The use of these judgements has been highlighted in the following areas:

- development of fault and event trees,
- potential lack of completeness,
- selection of plant damage states,
- selection of source terms,
- selection of applicable database,
- modelling dependent failures,
- modelling of human factors,
- modelling of physical processes,
- inclusion and modelling of hazards,
- methods for representing and propagating uncertainty.

In fact, there may be a need to adopt a deterministic approach in evaluating the contribution to risk due to human factors such as mis-diagnosis and malpractice.

We recommend that probabilistic safety assessments should contain quantitative estimates of the uncertainty associated with important outputs so far as reasonably practicable. This does not apply where a simple analysis shows clearly that targets are met.

10.3 Risk Evaluation

Chapter 6 shows that tolerance of risks can in principle be established in two ways: they might be shown to be below some threshold for concern; or they might be shown to be outweighed by the associated benefits. In both cases it is possible to look for useful reference levels of risk from risk comparisons.

Our principles for risk comparisons state that: different types of risk should be evaluated separately; voluntary components should be excluded; and qualitative differences between the risks to be compared should be identified and, if possible allowed for.

Although 'high consequence aversion' can in principle be applied to early deaths there is no satisfactory theoretical way to establish a numerical representation such as a CCDF shape. Cancer risks should in any case be treated simply in terms of the expected total number of deaths (subject to the associated uncertainty).

Lack of direct comparability of the hazards, and qualitative differences between the types of risk which do appear comparable, cause great *difficulties* for comprehensive, quantitative evaluation of

nuclear social risk against that of competing technologies for electricity production.

Reference levels of the social risk of cancer from nuclear accidents may be derived either from the existing cancer risk, or by comparison with background radiation.

However, when compared with the existing cancer risk, the number of effects would have to be very high to be detectable.

Comparison of the risk of early death from nuclear accidents with the existing accident risk is possible. However, it is not possible to construct from the data a CCDF which provides a useful reference level for risk evaluation.

The predicted risk from natural events may be helpful here.

Injury and birth defect risks from nuclear accidents can be managed as part of a strategy for managing the risk of early and delayed death respectively.

On the basis of historical data, we can find no useful reference levels for evacuation and non-personal risks.

10.4 Risk Management

Previous practice examined in Chapter 7 shows that as yet there is no universal agreement about the definition of tolerable risk targets. Furthermore, there is general unease about applying the ALARP principle; clear guidance is needed on the precise conditions to be satisfied. While there is considerable agreement that for the general public an individual risk of death of 10^{-6} per year is tolerable, there is lack of agreement about whether conditions that cause multiple deaths should be treated proportionately more restrictively. However, there is general support for the use of probabilistic methods for assessing safety, but less agreement about whether 'risk' is the appropriate form for targets. Although economic arguments have to be taken into account in making decisions about acceptability, there is no universal agreement about how such assessments should be made. No clear rationale for setting quantitative societal safety goals emerges from this international survey.

Chapter 8 found that formal decision theoretic techniques for balancing risks and benefits are beset with difficulties, though they do offer the advantage of clarifying the reasoning behind decisions. Public inquiry inspectors have considered societal risk to be important, but have struggled to interpret estimates usefully. They have shown no sign of high consequence aversion.

We also described so-called banded targets which consist of levels of risk which are high in some sense, levels which are low in some sense, and a requirement for the reduction of risk so far as practicable in the intervening region. Most existing quantitative risk management policies can be interpreted as complete or partial banded targets.

The use of banded targets is recommended so far as possible.

Formal decision theoretic techniques for balancing risks and benefits should not be mandatory.

However, if risk managers wish to use such a technique, a method to estimate the cost of risks to the health of the public is described in Appendix 1.

In using risk estimates to assess compliance with quantitative safety targets due account should be taken of the explicitly represented and implicit uncertainties associated with the estimates. It is not recommended that any particular mathematical technique to do this be mandatory.

10.5 The Overall Approach

Chapter 9 found that the fundamental social safety goal is that risks should be imposed only if the resulting benefits outweigh them. Because this is difficult to apply quantitatively, an alternative policy is to try to make the risks low enough to be of 'no concern,' and in any case to make sure they are not intolerable regardless of the benefit.

Therefore, risk management should in the first instance be based on levels of risk which can be shown to be low or high using reasoned arguments starting from reference levels of risk derived using appropriate principles.

Additional, more arbitrary, targets may be introduced on grounds which should be fully explained. Furthermore, because we recognise that risk management policies must be developed in an evolutionary way from current practice the primary targets may not always be phrased in terms of risk. Nonetheless the implied levels of risk should always be considered.

Risk managers should consider the use of a de minimis dose level, and, in any case, dose bands should be used to present the results of radiological risk estimates.

Risk estimates should be made on as realistic a basis as possible, taking into account the characteristics of the population at a specific site and the post-accident countermeasures which are likely to be taken. Estimates of financial harm should be undertaken appropriately.

Finally, it is desirable for the institutional provisions for public safety to be as clear to the public as possible, and for there to be greater public accessibility to safety cases and the dialogue between the organisations involved. Part of this need can be met by improving the presentation of: the needs of technicians and engineers; the risk estimation process; and the reasoning behind safety goals.

APPENDIX 1

COST BENEFIT ANALYSIS

A1.1 Introduction

Cost benefit analysis (CBA) is a technique which has been recommended as being helpful by the ICRP (1983) and the NRPB (1986) in making decisions about the degree of protection to be provided against radiation exposure. Proposals have also been put forward by the UKAEA (1987) in the recent Code of Practice. The recommended methods have been worked out for routine exposure and the purpose of this Appendix is to extend them to accident conditions.

Chapter 8 discusses the various different sets of options which may be considered using a formal decision aid such as CBA. The form recommended by the ICRP and the other proposals is that of a marginal analysis. The costs are those of putting into place and maintaining some protective system; the benefits are the financial and health benefits which accrue from this due to averted risk. The benefits of nuclear power are not factors in the argument because the available decision options do not include alternative (or no) energy sources.

In principle all the costs and benefits of taking some course of action must be considered to the extent that they differ between the various options. In planning for accident avoidance this means that the plant costs and other decontamination costs should be included in addition to those of the health detriment. It seems likely that the purely economic costs of disruption and decontamination will dominate the predicted costs of accidents. These can in principle be incorporated into CBA without difficulty, though in practice assessment methods may need to be developed, and the uncertainty will be large. The costs of the health detriment may be smaller than the others but they need to be considered properly because they should not be seen to be ignored. The remainder of this Appendix is thus concerned only with these health detriment aspects.

The previous recommendations for using CBA to control routine exposures are extended to accident conditions in section A1.2. Section A1.3 discusses the question of if and how to incorporate high consequence aversion.

A1.2 Application to Accident Conditions

It is often assumed that the routine operation and accident condition cases are completely different. However, they have similarities for the individual. As far as the individual is concerned there is no difference between his receiving 1 mSv per year or a chance of 100 mSv with a frequency of 0.01 per year. In either case his probability of developing a fatal cancer in each year of his life is the same; the deterministic nature of the routine exposure is lost through the stochastic nature of cancer induction.

The assumptions here are that the dose-risk relationship is linear and that the estimated probabilities of accidents can be treated on the same basis as the estimated probabilities of cancer induction.

On these grounds, if the societal risk were to be represented by the sum of all individual risks, the appropriate expectation value of collective dose could be used in the accident situation as a direct replacement for the collective dose from a routine case. It would then be appropriate to apply the routine man-sievert cost to the expected value of the collective dose. The routine exposure case has been extended to the accident situation by applying costs per unit of risk.

One complication needs to be considered: the NRPB (1986) costing scheme has a man-sievert costing which depends on annual individual dose, see Figure A1.1. Annual doses up to 1 Sv were considered: costs ranged from £3000 per manSv at the lowest level up to about £40,000 per manSv at the highest. The concept of cost per unit of risk can be applied to this figure by converting the horizontal axis of the NRPB curve to risk. It is then seen that the baseline value of £3000 is maintained up to risks of about 10^{-6} per year. This is the individual risk target for the public for UKAEA nuclear plant (UKAEA, 1987); if it is achieved individual risk aversion will have little effect. Worker individual risk may be higher than this and should be dealt with appropriately. In what follows we assume this effect to be negligible for accidents having a major social impact.

The NRPB recommendations apply only to the stochastic region, that is, we have been considering the risk of fatal cancer. Early deaths are usually discussed quite differently from late deaths. However, their effect on the risk for an individual can be treated in a very similar way. Although early death is a non-stochastic effect, the probabilistic nature of the accident means that we can discuss its impact in the same way as that of routine exposures and accidental exposures which do not cause early death. The only difference is that the risk of early death begins when the plant starts up; those of late death are delayed by the latent period for cancer induction. It is appropriate to deal with this effect in terms of loss of life expectancy. Table 6.1 shows that the loss of life expectancy from early death is about 40 years, whereas that from delayed death is 14 years on the models assumed there. While some factor could be introduced into the calculation to account for this (by weighting early deaths by a factor of 40/14 for example) it is unlikely to have a significant effect in relation to the other uncertainties. From now on we ignore this difference.

If we have an accident with frequency f, the expectation value of its economic cost through health effects over the plant lifetime, T_P, is

$$C = fT_P(S_pQ + N_E)(M/Q)$$

where S_p is the collective dose, M is the baseline value of the man-sievert, N_E is the number of early deaths resulting and Q is the risk of death per unit dose introduced in Chapter 3, with a value of about 10^{-2} per Sv.

The constant (M/Q) is the cost per unit of risk, and can thus also be considered as the cost of a life. It is about £300,000. This is towards the low end of the range usually considered for this figure. The UKAEA (1987) Code of Practice allows for manSv costings ten times higher than the NRPB baseline when all relevant factors are taken into account (see Chapter 7), and this might also be reasonable for accidents.

Finally we note that all relevant factors should be taken into account in design for accident avoidance. Thus doses which would be received as a result of cleanup operations from a hypothetical accident should, if possible, be considered in addition to those received involuntarily from the accident itself.

A1.3 High Consequence Aversion

To the extent that the consequences of an accident can be measured by the sum of its effect on individuals, the above discussion suggests that the valuation of radiation exposures can be derived directly from values for routine exposure; indeed the same value should be used for the man-sievert and the 'expected man-sievert'. However, in considering societal risk it is frequently argued that the public attitude to accidental exposures implies that it is not sufficient to consider just the exposure of individuals. This is because of 'risk aversion', or more strictly 'high consequence aversion': society is more concerned about events which cause a large number of simultaneous deaths than several events which cause, in total, the same number. If this effect is to be included in a CBA approach to societal safety it is necessary to modify the treatment described above.

In terms of consequence aversion there is no need for late deaths and early deaths to be treated in the same way. Late deaths would not occur simultaneously but over the following 30 years or more. Early deaths would occur generally within one year. Thus, it might be argued that the aversion to early deaths would be the greater; it would not be an unreasonable starting point to assume no consequence aversion to late deaths.

The mathematical expression of consequence aversion could take several forms. However, it is plausible to regard it as depending only on the total consequence and not the distribution of individual risk. In this case we can write the societal cost of the accident as

$$C = fT_P(S_pQ + N_E)(M/Q)A(S_p,N_E)$$

where A is the consequence aversion factor. If we assume no consequence aversion to late deaths

$$C = fT_P(S_pQ + N_E)(M/Q)A(N_E) \qquad (A1.1)$$

with $A(1)=1$ and A increasing with N_E.

An alternative form of this would apply consequence aversion only to the early deaths term:

$$C = fT_P(S_pQ + N_EA(N_E))(M/Q). \qquad (A1.2)$$

The two forms would reflect different reactions on the part of society to the accident. In the first case it is assumed that the early deaths draw society's attention disproportionately to all deaths, whereas in the second case it is to early deaths alone.

In order to proceed it is necessary to specify a satisfactory form for $A(N_E)$. There are of course many which fit the conditions so far defined. In choosing one it appears sensible to take account of the considerable uncertainties in the nature of consequence aversion and, indeed, in the calculations which are to be made in applying CBA (where, for example, f may be uncertain to within a few decades). Thus the form chosen should be simple with as few adjustable parameters as possible. Following previous work it is suggested that a simple power law is adequate:

$$A(N_E) = N_E^a \quad \text{with} \quad a>0.$$

Selection of the power a is difficult. Intuitively it seems that it should be less than one: to spend 10^4 times more to avoid 100 deaths rather than one death seems unreasonable. For no consequence aversion $a=0$.

The cost associated with given numbers of deaths are shown in Table A1.1. The plant lifetime, T_P, has been taken as 30 years and the man-sievert value, M, as £3000. The frequency of the accident has been taken at 10^{-7} per year; the accident therefore would contribute about 1% of an individual risk target for a member of the public of 10^{-6} per year. (On the assumption that, given a fatal accident, the risk of death to the most exposed member of the public is about 10%, the chance that the wind is blowing in his direction.)

The figures indicate that, even with severe risk aversion, it is possible to generate only £10^6 from health effects costs when more than 10^4 early deaths are predicted or 10^6 late deaths. This refers to a frequency of 10^{-7} per year; it the frequency is lower, the health cost of the accident falls proportionately.

The power law form of the consequence aversion factor is similar to the power laws that are frequently assumed for CCDF targets. However, CBA is a quite separate matter just as, in routine exposure cases, it is distinct from the requirement to comply with dose limits. This is explained further in Chapter 8.

A1.4 Conclusions

(1) Cost benefit analysis can be extended from the routine exposure case, as recommended by the ICRP and NRPB, to the accident case by working in terms of the cost per unit of imposed risk.

(2) Consequence aversion can be modelled in principle, but there is no clearcut way in which this should be done.

(3) It seems unlikely that the health detriment costs from accidents would be significant in such an analysis given that the other currently existing safety targets are met.

A1.5 References

ICRP, 1983, **Cost Benefit Analysis in the Optimisation of Radiation Protection**. ICRP Publication 37, Ann ICRP, 10, No 2/3.

NRPB, 1986, **Cost Benefit Analysis in the Optimisation of Radiological Protection**. ASP 9, HMSO, London.

UKAEA, 1987, **Code of Practice and Guidance Note: Radiological Guidelines for the Design and Operation of UKAEA Plant.** Safety and Reliability Directorate Report, SRD R 456.

TABLE A1.1
Societal Costs of Accidents

The costs in (a) and (b) are given in pounds

(a) Consequence aversion applied as in equation A1.1

N_E	a	Number of late deaths		
		10^2	10^4	10^6
1	0	9.1×10^1	9.0×10^3	9.0×10^5
	0.2	9.1×10^1	9.0×10^3	9.0×10^5
	0.5	9.1×10^1	9.0×10^3	9.0×10^5
10^2	0	1.8×10^2	9.1×10^3	9.0×10^5
	0.2	4.5×10^2	2.3×10^4	2.3×10^6
	0.5	1.8×10^3	9.1×10^4	9.0×10^6
10^4	0		1.8×10^4	
	0.2		1.1×10^5	
	0.5		1.8×10^6	

(b) Consequence aversion applied as in equation A1.2

N_E	a	Number of late deaths		
		10^2	10^4	10^6
1	0	9.1×10^1	9.0×10^3	9.0×10^5
	0.2	9.1×10^1	9.0×10^3	9.0×10^5
	0.5	9.1×10^1	9.0×10^3	9.0×10^5
10^2	0	1.8×10^2	9.1×10^3	9.0×10^5
	0.2	3.2×10^2	9.2×10^3	9.0×10^5
	0.5	9.9×10^2	9.9×10^3	9.0×10^5
10^4			1.8×10^4	
	0.2		6.6×10^4	
	0.5		9.1×10^5	

FIGURE A1.1
Multiplier to be Applied to Baseline Detriment Costs
(Source: NRPB, 1986)

Multiplier to be applied to baseline detriment costs (£3000 per manSv) as a function of annual individual dose. The uncertainty is indicated by the fuzziness of the line. The NRPB emphasise that factors other than the individual risk aversion shown here may be relevant, for example, proximity to dose limits, uncertainty in dose calculations, the degree to which the exposure is voluntary, or even political considerations.

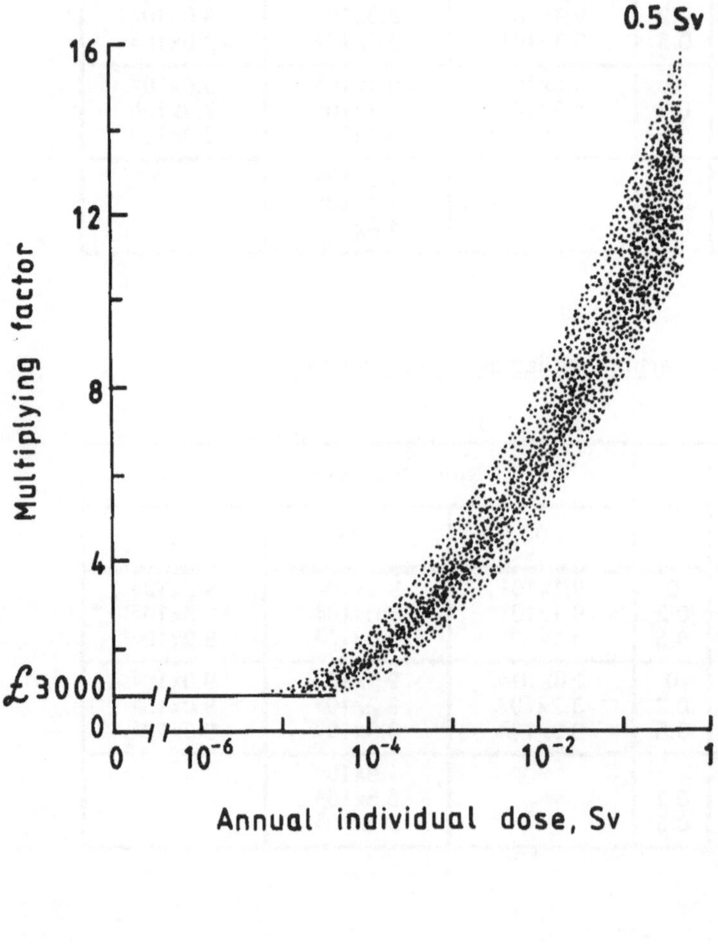

APPENDIX 2

GLOSSARY AND ACRONYMS

References to other defined terms are underlined.

A2.1 Glossary

absorbed dose (or dose) amount of energy deposited per unit mass by ionising radiation. Measured in grays (Gy) or rads.

activity rate at which transitions take place in a radionuclide, and hence the rate of emission of radiation. Measured in becquerels (Bq) or curies.

ALARA see ALARP

ALARP the basic requirement for risk management in the UK. The legal definition requires a calculation in which the benefits **grossly** outweigh the risks, but the way in which each is to be assessed is not clearly defined by Government. ALARA, as required by the ICRP does not require gross disproportion and it is increasingly accepted that it can be implemented in part by CBA.

background risk existing levels of risk (usually calculated by direct estimation) against which the risk levels of a new development can be compared.

banded targets risk targets formulated in terms of levels of risk which are intolerable regardless of the benefit, levels which are de minimis together with a requirement to reduce risks as far as practicable (or some similar concept) in between.

Bayesian methods a means of statistical inference which depends on assigning probability distributions to the parameters to be estimated.

best estimates the use of data and models in a safety case which are not deliberately pessimistic or conservative.

beyond design basis analysis analysis of faults which the plant may not be able to withstand in some defined way. See design basis analysis.

binning see categorisation

categorisation a PSA technique whereby accident sequences are grouped together on the basis of similarity of certain properties in order to reduce calculational effort. Also known as binning or pinch pointing.

CIMAH regulations the UK response to the Seveso directive which set out the safety requirements to be satisfied by plant with the potential for major accidents.

collective effective dose equivalent sum of effective dose equivalent over a population. Measured in man-sieverts (manSv) or man-rems (not to be confused with millisieverts, mSv, or millirems, mrem).

collective effective dose equivalent commitment the total collective effective dose equivalent arising from a particular source of radiation over a period of time in which the population may change.

committed effective dose equivalent the effective dose equivalent over 50 years arising from the incorporation of radioactivity into the body. Units same as effective dose equivalent.

common cause failure a dependent failure where simultaneous multiple failures result from a single shared cause. See also common mode failure.

common mode failure a common cause failure in which each item fails in the same mode.

complementary cumulative distribution function a graph in which the horizontal axis is the size of a consequence and the vertical axis is the frequency or probability with which each consequence level is equalled or exceeded. Most common means of representing social risk.

completeness problem the question of whether all significant events have been accounted for in a PSA.

consequence analysis the modelling of the consequences of accidents starting with a source term.

consequences the undesired events associated with risk which may be human injury, damage to property or damage to the environment (see hazard (1)).

conservative making pessimistic assumptions in a safety analysis in order to be assured of erring on the safe side. See best estimates.

containment analysis analysis of the physical behaviour during accidents of the various barriers to release of radioactivity.

cost benefit analysis a variant of decision theory in which all the factors relevant to a decision are measured in financial units. See also cost effectiveness.

cost effectiveness a variant of CBA which compares value for money of various options.

countermeasures measures such as evacuation, interdiction of land and food or decontamination taken to minimise the consequences of an accident.

de minimis of no account or trivial.

decision theory a mathematical means of weighing various options in order to reach a decision as to which is best.

dependent failure the situation where causal links mean that failure of one component is not independent of failure of others. See also common cause failure.

design basis analysis analysis of faults which a plant must be designed to withstand. Uses prescribed, conservative methods. See beyond design basis analysis.

Design Safety Criteria/Guidelines the CEGB's requirements for the suppliers of nuclear reactors. Inter alia, the criteria contain general reliability targets, whereas the Guidelines expand the criteria for specific reactor types to set reliability targets for each safety system.

deterministic approach general term for non-probabilistic approaches to safety, but the precise definition may vary widely.

direct estimation the estimation of risks directly from historical data on the occurrence of the events of interest, in contrast to indirect estimation which requires a modelling technique such as PSA.

dose see absorbed dose, but also used loosely for any of the other dose terms.

dose equivalent dose multiplied by a quality factor to account for the differing biological effects of different types of radiation. Measured in sieverts (Sv) or rems.

effective dose equivalent the dose equivalent to each organ is multiplied by weighting factors and summed to give a number which represents the harm from stochastic health effects in terms of a whole body dose equivalent. Same units as dose equivalent.

emergency reference level the dose equivalent (for various organs) at which countermeasures should be considered or taken.

engineering judgement the process whereby informed opinion is used to make progress in situations where wholly objective methods are not available.

event tree a logic model in which the various possibilities stemming from a starting event are enumerated.

expected utility theory a decision theory method which relies on measuring the outcomes on a single (utility) scale and taking the option which maximises expected utility under a suitably assigned set of probabilities.

external events see hazards (2)

F-N line same as CCDF.

Farmer line a risk target in which frequencies are set for individual accidents as a function of their release in terms of equivalent I-131 activity.

fault tree logic model in which the combinations of faults leading to an overall fault (called the top event) are enumerated.

fission product transport the analysis of how fission products are transported in a reactor accident with given containment behaviour, as determined from containment analysis), starting with a given plant damage state and resulting in a source term.

frequency the expected number of events per unit time which, for a Poisson process (the usual PSA statistical model), is equal to the Poisson parameter. Also used by statisticians for the number of specified events in a sample. This is different.

global analysis application of decision theory in which all options are considered, as distinct from marginal analysis.

hazard (1) a physical situation with a potential for human injury, damage to property, damage to the environment or some combination of these.

hazard (2) a PSA term for an initiating event which does not result from the plant transients or failures, for example, fires and earthquakes. Also known as external events.

individual risk the frequency at which an individual may be expected to sustain a given level of harm from the realisation of specified hazards (1).

Level 1 PSA a PSA which analyses initiating events up to the stage of the resultant plant damage states. Essentially a reliability analysis.

Level 2 PSA a PSA which analyses faults up to the stage of specifying releases or source terms.

Level 3 PSA a PSA which analyses faults as far as the final consequences.

life expectancy the average age at death under a specific set of conditions minus age at present. Thus it is a function of present age.

logic models generic term for system analytic methods such as fault trees and event trees.

marginal analysis application of decision theory which examines a restricted set of choices as distinct from global analysis, generally those which accept that an activity will be carried out and the only options reflect how (provision of safety features, for example).

mistakes cognitive human errors (as opposed to slips): "if the intention is not appropriate, this is a mistake."

objective not containing any element of judgement, as opposed to subjective.

phenomenology term for the physical, chemical, biological and psychological models that are used in PSA.

plant analysis same as Level 1 PSA.

plant damage states PSA term for the chosen set of outcomes of plant system failure. For example, modes of core melt.

probabilistic risk/safety assessment the modelling of accidents on a plant so as to estimate the frequency of various undesired events.

probability (of an event) the long run fraction of realisations of a particular situation which result in the event.

radioactivity the property of spontaneously emitting ionising radiation. Also material having this property.

Rasmussen report see Reactor Safety Study

Reactor Safety Study first major PSA of nuclear reactors. Produced in the US in 1975.

risk the likelihood of specified undesired events occurring within a specified period or in specified circumstances from the realisation of a specified hazard (1). It may be expressed as either a frequency or a probability depending on the circumstances.

risk acceptability see tolerable risk

risk assessment the general term used to describe the study of decisions subject to uncertain consequences; the combination of risk estimation and risk evaluation.

risk aversion this has both a general and a technical usage. Its technical use in decision theory is for convex utility functions; for example disproportionately higher valuations of larger undesired consequences. This is called 'high consequence aversion' in this report.

risk estimation general term for the process of estimating risk which includes: the identification of the outcomes; the estimation of the magnitude of the associated consequences of these outcomes; and the estimation of the probabilities of these outcomes.

risk evaluation the complex process of determining the significance or value of the identified hazards (1) to those concerned with or affected by a risk management decision.

risk management the making of decisions concerning risks. Flows from risk estimation and risk evaluation.

risk perception the evaluation of risks by individuals in society in non-quantitative terms.

Safety Assessment Principles the guidance given by HMNII to its inspectors.

Seveso directive EEC directive to member states on implementing a policy for managing major chemical hazards following the accident at Seveso.

slips human error in carrying out a procedure (as opposed to mistakes): "if the action is not what was intended, this is a slip."

social risk the frequencies with which specified numbers of people in a given population, or the population as a whole, sustain a specified level of harm from the realisation of specified hazards (1).

societal risk see social risk

source terms PSA term for the chosen set of releases. Specified in terms of amounts of radionuclides, associated materials, temperature, release height and so on.

stochastic health effects (generally following radiation exposure) those for which the severity does not depend on exposure, though the probability of developing the effect does.

subjective containing some element of judgement as opposed to objective. This is generally used in the context of risk estimation where it is necessary to make some judgement in order to assign numerical values and hence make progress with the quantitative method. Subjective probability is an inherent constituent of Bayesian methods and decision making using expected utility theory.

success criteria PSA term for the assessed minimum performance level for a system to avoid a some specified accident condition. Usually forms a fault tree top event and is determined from transient analysis.

tolerable risk a risk that society as a whole (though not necessarily every member) chooses to accept because of the associated benefits. This does not mean it is an 'acceptable' risk.

transient analysis PSA term for the analysis of time dependent events (either during normal operation or during accidents) in nuclear reactors.

Wash-1400 see Reactor Safety Study

A2.2 List of Acronyms

ACMH	Advisory Committee on Major Hazards
ALARP	as low as reasonably practical
ALARA	as low as reasonably achievable
ALATA	as low as technically achievable
ARLs	assessment reference levels (of the SAPs)
BNFL	British Nuclear Fuels plc
BWR	boiling water reactor
CBA	cost benefit analysis
CCDF	cumulative complementary distribution function
CEGB	Central Electricity Generating Board
CIMAH	Control of Major Accident Hazard (Regulations)
DSGs	Design Safety Guidelines of the CEGB
EDRP	European Demonstration Reprocessing Plant
ERL	Emergency Reference Level
FAR	fatal accident rate
FMCD	first moment cumulative distribution
FMEA	failure modes and effects analysis
HAZOPS	hazard and operability studies
HMNII	Her Majesty's Nuclear Installations Inspectorate
HSC	Health and Safety Commission
HSE	Health and Safety Executive
IChemE	Institution of Chemical Engineers
ICRP	International Commission for Radiological Protection
JIC	(Orkney and Shetlands) Joint Islands Council
LLE	loss of life expectancy
LPG	liquefied petroleum gas
LWR	light water reactor
NE	Nuclear Electric
NGL	natural gas liquids
NII	Nuclear Installations Inspectorate
NRPB	National Radiological Protection Board
PRA	probabilistic risk assessment
PSA	probabilistic safety assessment
PWR	pressurised water reactor
QALY	quality adjusted life year
RSS	Reactor Safety Study
SAPs	Safety Assessment Principles of the NII
SNG	substitute natural gas
SRD	Safety and Reliability Directorate
TMI	Three Mile Island
UKAEA	United Kingdom Atomic Energy Authority (trading name AEA Technology)
USNRC	United States Nuclear Regulatory Commission
WGRSPMA	Working Group on the Risks to Society from Potential Major Accidents